# HOT SPRING
# RESORT & SPA

温泉度假村

赵欣 译

高迪国际出版有限公司 编
大连理工大学出版社

图书在版编目 (CIP) 数据

温泉度假村：汉英对照 / 高迪国际出版有限公司编
；赵欣译. —大连：大连理工大学出版社，2013.2
　ISBN 978-7-5611-7611-5

Ⅰ．①温… Ⅱ．①高…②赵… Ⅲ．①温泉—旅游度假村—建筑设计—世界—图集 Ⅳ．① TU247-64

中国版本图书馆 CIP 数据核字 (2013) 第 020527 号

出版发行：大连理工大学出版社
　　　　　（地址：大连市软件园路 80 号 邮编：116023）
印　　刷：利丰雅高印刷（深圳）有限公司
幅面尺寸：245mm×320mm
印　　张：21
插　　页：4
出版时间：2013 年 2 月第 1 版
印刷时间：2013 年 2 月第 1 次印刷
策划编辑：袁　斌　刘　蓉
责任编辑：刘　蓉
责任校对：李　雪
封面设计：郭　聪

ISBN 978-7-5611-7611-5
定　　价：320.00 元

电话：0411-84708842
传真：0411-84701466
邮购：0411-84703636
E-mail:designbooks_dutp@yahoo.cn
URL:http://www.dutp.cn

如有质量问题请联系出版中心：（0411）84709246　84709043

# PREFACE
## 序言

"SPA, hot spring and bath" are a common leisure activity at all times. It used to be a ritual, a trend, a health preservation method, and a luxurious enjoyment. Tangquan Palace is considered as a treasure place by the rulers of Zhou, Qin, Han, Sui and Tang Dynastyies. The Huaqing Hot Spring and the good portrayal "bathing in Huaqing Hot Spring in cold spring and enjoying the water touching the skin" are well-knowned by the Chinese. In the modern time of economic prosperity, people have an affluent life, thus this ancient and luxurious activity becomes a necessary facility flaunted by various hotels and resorts.

People need to take a break in the busy life. They can ease the physical fatigue through sleeping, but the mental tiredness should be wiped off through leisure activities, for example, listening to the music can soothe the mind, reading can cultivate taste, and traveling can relax the tense mood... the most special one is hot spring, which can not only help them recover from fatigue, but also endow them with joyful soul. They take a hot spring bath of mineral and enjoy the beautiful scenery in front, and the weary soul gradually relaxes with the stretch of body, and the warm hot spring water interacts between the pores and cortical cells. The sensory feeling allows their body and the spring to further communicate, and finally they enjoy not only a physical rest, but also a spiritual satisfaction.

At first, taking a hot spring bath is just to warm the body to against the cold, but later people gradually discovered the effect of hot spring on treatment. With the improvement of living standard, hot spring bath, along with visiting folk culture and enjoying local specialties, has become spa tourism activities that people are familiar with, which is a leisure activity integrating with body and mind. I like a tourism slogan very much, it says setting out to travel is not only to repeat oneself, but to try to feel the surrounding with attention, starting once again. When I went to take hot spring bath, I took my family or some friends, and we chose morning, evening or night to enjoy the sunrise, sunset or starlight; hearing the frogs calling and smelling the sweet-scented osmanthus, and we can feel our mood starting once again. The most profound memory is that we went to a hot spring beside a wild rivulet on the hill at night. It is a pool piled with stone on site by people camping out, in the river bed, and cold water from the stream flew and splashed on our bodies. In the pitch-dark night, we occasionally took out a flashlight and could shine on the owl in the branches, and the gurgling streams and our hip-hop sound of laughter form an impressive and beautiful screen. I cannot forget the special feeling forever.

The activity of hot spring bath features various ways, functions and forms around the world, but I think hot spring bath should not be "fishing in rushing water", but to allow people to enjoy a close and intimate contact with nature through hot spring. Warming body in hot spring, we change the beautiful landscape, nice music and good friends into good ingredients and put them into the "soup" to soak slowly, and our bodies and minds are integrated with the hot spring into a whole. With physical stretch and spiritual washing, we can feel the gift of nature carefully, so as to achieve the combination of mind and nature.

LAI LIAN QU
HANCS Landscape Planning Co.,Ltd.
December 19, 2012

"SPA、温泉、沐浴"自古以来就是一项共同的休闲活动。它曾经是一种仪式、一种风潮、一种养生的技术、一种奢华的享受。古有周、秦、汉、隋、唐历代统治者都视为宝地的"汤泉宫",我们熟知的"华清池"、"春寒赐浴华清池,温泉水滑洗凝脂"便是很好的写照。而在经济繁荣人民生活富足的现代,这项古老而奢华的活动则成为各大酒店、度假村极力追捧的必备设施。

在忙碌之余,人们需要休息。身体的疲惫可以通过睡眠来消除,而心灵的休息则靠休闲活动来调节,比如听音乐可以舒缓心灵,看书可以陶冶性情,旅游可以放松紧张的心情……其中最为特别的一项当属泡温泉,它既能解乏又能愉悦心灵。一边泡着含有矿物质的泉水,一边欣赏眼前美丽的风景,疲乏的心灵随着舒展的身体逐渐放松,暖热的泉水在毛孔及皮层细胞间交互作用,感官带动身体与泉水进一步"交流沟通",最终,人们得到的不仅是身体上的休息,更是心灵上的满足。

泡温泉最初只是为了温暖身体、抵御严寒,后来人们才逐渐发现泡温泉在理疗上的效果。随着生活条件的改善,泡温泉和参观民俗文化、吃当地特色美食一同成为人们熟悉的温泉旅游活动,也就是结合了身心为一体的休闲活动。有一句旅游广告语我很喜欢:出发去旅行不是为了再一次重复自己,而是尝试用心灵感受周围,再一次出发。我每一次去泡温泉时都带着家人或者几个友人,选择早上、黄昏或夜里,可以看到日出、日暮或星光;听到蛙鸣,闻到桂花香,感觉都是心情的再一次出发。记忆最为深刻的一次就是夜里到山上野溪旁泡温泉。那是野营的人们用现场的石头堆砌的泡池,泡池就在溪床上,不时有冷水从溪边流过溅到身上,漆黑的夜里偶尔用手电筒还可以照到停在树枝上的猫头鹰,潺潺溪流声还有我们一伙人嘻哈的笑语声,形成了一幅有声有色的美丽画面,让我永远无法忘怀。

泡温泉这项活动在全世界有着各种方式、功能、形态,但我认为泡温泉不是在汤汤水水中"浑水摸鱼",而是让人们的身体通过温泉零距离地与自然亲密接触。用泉水温暖身体,将身边的美景、美乐、好友变成好料放进"汤"中慢慢泡,身心与泉水交融为一体,身体得到舒展,心灵得到洗涤,用心感受大自然的恩赐,从而达到心灵与大自然的合二为一。

HANCS Landscape Planning Co., Ltd.
2012年12月19日

# CONTENTS
目 录

**006** Mingtang Hot Spring Resort
茗汤温泉度假村

**118** Fujian Strait Culture Homeland Hot Spring Resort
福建海峡文化原乡温泉度假村

Longmen DiPai Hot Spring Resort
龙门地派温泉酒店

**132** Global Villa Hot Spring Paradise
锦绣香江温泉城

**062** Wuxi Lingshan Yuanyi Ri-Star Spa Hotel
无锡灵山元一丽星温泉度假酒店

**142** Zhongshan Hot Spring Hotel
中山温泉宾馆园林改造

Mineral Garden and Spa
矿泉花园温泉

**152** Tianmu Lake Tuanbo Holiday Hotel
天沐湖团泊假日酒店

Dino-valley Hot Spring Resort Changzhou
常州恐龙谷温泉酒店

Tianyi Hot Spring Resort
天一温泉度假村

 182 One & Only Hotel
开普敦 One & Only 度假酒店

 256 Qingyuan Shampoola Resort Hotel
清远森波拉度假酒店

 202 Hot Spring Leisure City
温都水城

 270 Nanchang Jinyan International Hot Spring Resort Island
南昌金燕国际温泉度假岛

 210 Wuxi Lingshan Yuanyi International Resort
无锡灵山元一国际度假村

 278 Nanhu Travel Phoenix Lake International Hot Spring Resort
南湖国旅凤凰湖国际温泉度假村

 218 A Garda Lake Hotel with Ayurvedic Style
加尔达湖阿育吠陀风格酒店

 288 Riyuegǔ Hot Spring Country Club, Phase II
日月谷温泉乡村俱乐部二期

 230 Lintong · Aegean International Hot Spring Hotel
临潼·爱琴海国际温泉酒店

 306 Yuanchang Yinhe Central Lake Hot Spring Resort
源昌银河湖心岛温泉度假村

 246 Nanjing Yuhao Tangshan Hot Spring International Hotel
南京御豪汤山温泉国际酒店

 312 Huidong Baipenzhu Hot Spring Hotel
惠东白盆珠温泉酒店

334 INDEX 索引

# Mingtang Hot Spring Resort

茗汤温泉度假村

| | |
|---|---|
| 项目地点 | 河北霸州 |
| 项目面积 | 120000平方米（第一期）|
| 会馆面积 | 13000平方米 |
| 设 计 师 | 陈源盛、曾锦玲、陈亚馨、Annelie Hakansson |
| 设计公司 | 城市设计联盟 |
| 景观设计 | LPD Landscape Planning & Design Inc |
| 照明设计 | MICHIKO YOKOTA and LIGHT |
| 客　　户 | 茗汤休闲度假集团 |
| 摄　　影 | 小雄梁彦影像有限公司 |

How to redefine quality for 5 star resort hotel which is not mainly by it's physical luxury but more about rich nature experience is the issue the designers have and how to create resort as one sustainable environment is their main concern. Therefore, environment goes first, then landscape experience, architectural form is the last step. For them, architectural form is just based on how to integrate it with surroundings.

More Nature, More Sustainable

New topography created intends to respond to site's situation which is to have high hill at north side working as defense for cold winter wind from the northwest and low hill at south side to guide cool summer wind from the southeast into the site. Topography also works as base to creates different landscape typologies which will create multiple landscape experience — the hilltop as grassland, the hillside as forest, and the low land as hot spring. Hotel and villas will be set at different position and zoning as groups with different landscape themes such as hot spring resort, forest cottage, and house by the lake.

Architecture in Nature, Nature in Architecture

"Architecture in Nature, Nature in Architecture" is the basic concept for the hotel. In the way, the building is not the one that concentrates on form and style; they intend to create weak and humble architecture which is harmonious and consistent with surroundings. They take linear form for building and the linear form is as a loop. It has been put on topography which still will keep landscape inside and can give maximum proximity and access to the landscape outside. Moreover, it also works as a three-dimensional corridor to guide guests experiencing environment and as the best circulation for flow of people as well.

The designers hope this resort could be an environment which people can experience by not just only vision, but hearing, smell, and touch as well. It is just simply you can hear the voice of water, birds, and wind going through bamboo, and can feel the hot spring and smell the flower as well.

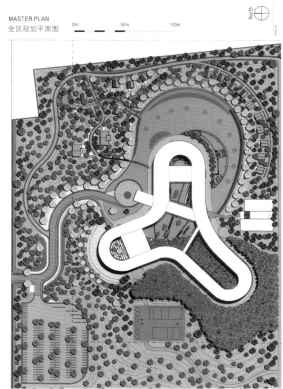

| HOTEL MODEL 旅馆建筑模型 | | | 1. Hotel entrance 温泉旅馆入口广场 | LANDSCAPE MODEL OF COURTYARD 景观中庭模型 | | |
|---|---|---|---|---|---|---|
| A. WATER COURTYARD | 水景中庭 | | 2. Lobby 入口大厅 | A. WATER COURTYARD | 水景中庭 |
| B. WATER COURTYARD | 水景中庭 | | 3. café 温咖啡厅廊道 | B. WATER COURTYARD | 水景中庭 |
| C. BAMBOO COURTYARD | 竹林中庭 | | 4. Bridge 景观桥 | C. BAMBOO COURTYARD | 竹林中庭 |
| D. FLOWER COURTYARD | 乾景中庭 | | 5. Restaurant 餐厅开放区 | D. FLOWER COURTYARD | 乾景中庭 |
| E. WATERFALL COURTYARD | 水瀑中庭 | | 6. Restaurant room 餐厅包厢区 | E. WATERFALL COURTYARD | 水瀑中庭 |
| | | | 7. Indoor swimming pool 室内温泉泳池区 | | |
| | | | 8. outdoor swimming pool 户外温泉泳池区 | | |
| | | | 9. SPA | | |
| | | | 10. 2nd Guest room & Corridor 二楼客房区及廊道 | | |

1

traditional block hotel break into small room
superior to the surrounding

2

by bending the line
we optimize with the landscape connection

3

the proximity to the clubhouse and restaurant facilities

4

landscape site plan

5

6

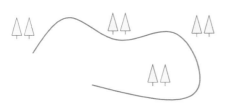

7

circulation scheme

8

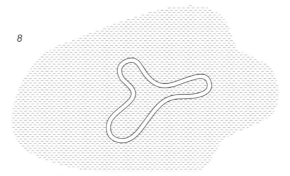

by creating a low density building gently layed out over
the hotspring landscape we aviod creating a very heavy
solid builing and instead the building keeps the landscape
on all sides of the facades.

9

Mingtang hotspring resort

如何通过环境带给人心灵感受上的丰富性，而不是根据豪华程度来定义五星级度假旅馆的质量是设计师关心的议题。因此，此案的目的首先是要塑造一个适合的环境，然后再置入景观的主题与经验，最后才是建筑形式的考虑。对设计师而言，建筑最后的形式只是在思考如何与周围形成的环境相融合。

**更多的自然，更多的永续**

在全区布局上，藉由重新塑造的地形来响应气候对基地的影响：在北面，较高的地形如同屏障一般来阻挡冬季寒冷的西北季风，在南面则以低的地形引导夏季凉爽的东南季风进入基地。而不同的地形也形成了不同的景观类型及塑造了多样性的景观经验：丘顶 - 草原／山坡 - 树林／低地 - 温泉。环境形成之后，在不同的地形置入会馆及别墅，形成了不同景观主题的群落，例如温泉会馆、湖畔小屋或是林中木屋。

**自然中的建筑，建筑中的自然**

自然中的建筑、建筑中的自然，是我们会馆规划的主要概念。简而言之，它并不是一个强调造型及风格的建筑。设计师希望塑造一个谦虚且能跟环境融合一致的空间。建筑有如一个线性的环状量体，置入到基地里面却并不会将环境盖掉，同时还能带给空间与景观最大的接近机会。设计师希望建筑空间本身就如同一个三维的循环动线，是一条能够将人带入及体验环境的廊道。

设计师希望这个度假村是一个可以让人在视觉、听觉、嗅觉及触觉上都能够感受得到的环境；能听到水声、鸟声及风吹过竹林的声音，能感受到温泉及雾气，也能闻到花的味道。

温泉会馆立面图 ELEVATION

 Stone 石材

 Wood 实木

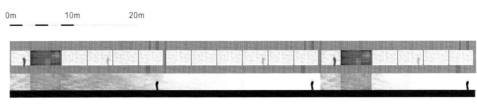

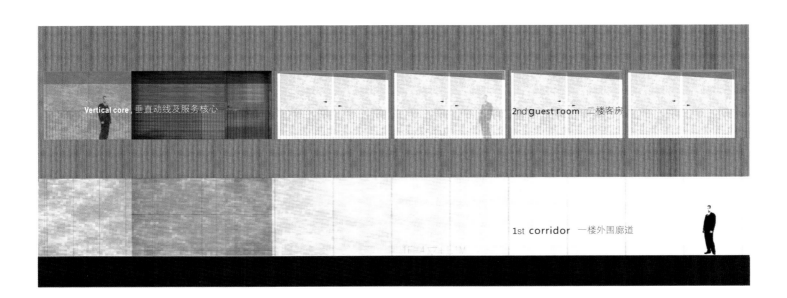

Vertical core 垂直动线及服务核心

2nd guest room 二楼客房

1st corridor 一楼外围廊道

### 建筑 & 景观 剖面图一  SECTION-1

景观桥 - 水景中庭 - 大厅区
Bridge - Water Courtyard - Lobby

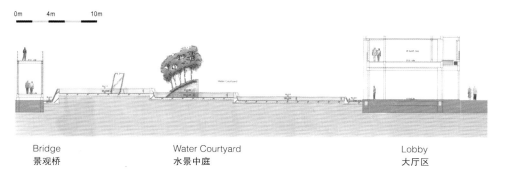

Bridge
景观桥

Water Courtyard
水景中庭

Lobby
大厅区

### 建筑 & 景观 剖面图二  SECTION-2

室内温泉泳池区 - 水瀑中庭 - SPA区 - 户外温泉泳池区 - 景观湖
Indoor Swimming Pool - Waterfall Courtyard - SPA - Outdoor Swimming Pool - Lake

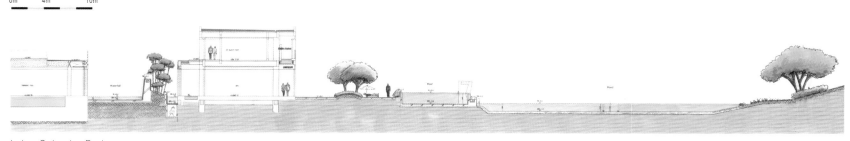

Indoor Swimming Pool
室内温泉泳池区

Waterfall Courtyard
水瀑中庭

SPA区

Outdoor Swimming Pool
户外温泉泳池区

Lake
景观湖

Path
湖边步道

### 建筑 & 景观 剖面图三  SECTION-3

餐厅包厢区 – 乾景中庭 – 餐厅用餐区 – 水景中庭
Restaurant - Flower Courtyard - Restuarant - Water Courtyard

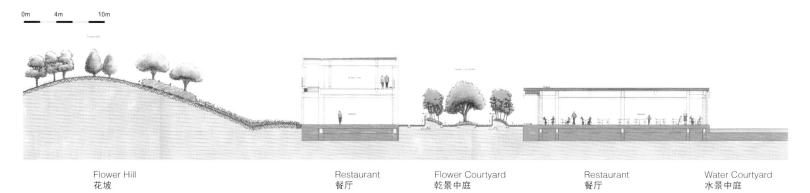

| Flower Hill | Restaurant | Flower Courtyard | Restaurant | Water Courtyard |
| 花坡 | 餐厅 | 乾景中庭 | 餐厅 | 水景中庭 |

### 建筑 & 景观 剖面图四  SECTION-4

客房区 – 竹林中庭 – 客房区
Guest Room - Bamboo Courtyard - Guest Room

Bamboo Hill　竹林　　Guest Room　客房　　Bamboo Courtyard　竹林中庭　　Guest Room　客房区

温泉会馆一楼平面图 1F PLAN

0m　　　50m

1. 主入口 Entrance
2. 入口大厅 Lobby
3. 咖啡廊道 Cafe
4. 商店 Shop
5. 餐厅开放区 Restaurant
6. 餐厅包厢区 Restaurant
7. 酒吧 Bar
8. 厨房 Kitchen
9. 会所大厅 Lobby
10. 男更衣室 M Locker Room
11. 女更衣室 F Locker Room
12. 室内泳池区 Indoor Swimming Pool
13. SPA
14. 健身房 Gym
15. 景观桥 Bridge
16. 竹林客房区 Guest Room

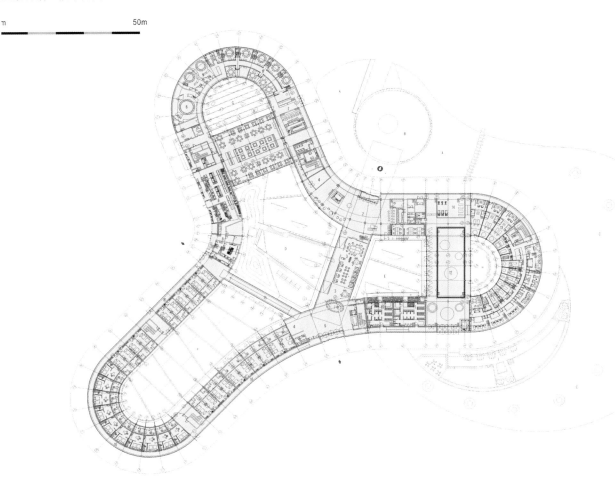

A. 景观池 Pool
B. 水广场 Water Plaza
C. 生态湖 Lake
D. 水景中庭 Water Courtyard
E. 水景中庭 Water Courtyard
F. 竹林中庭 Bamboo Courtyard
G. 乾景中庭 Flower Courtyard
H. 水瀑中庭 Waterfall Courtyard
I. 户外泳池区 Outdoor Swimming Pool

温泉会馆二楼平面图 2F PLAN

0m　　　50m

1. 景观露台 Terrace
2. 景观桥 Bridge
3. 景观阳台 Terrace
4. 水瀑 Waterfall
5. 二楼大厅 Lobby
6. 梯厅 Small Lobby
7. 客房廊道 Corridor
8. 客房（标间）Guest Room
9. 客房（套间）Guest Room

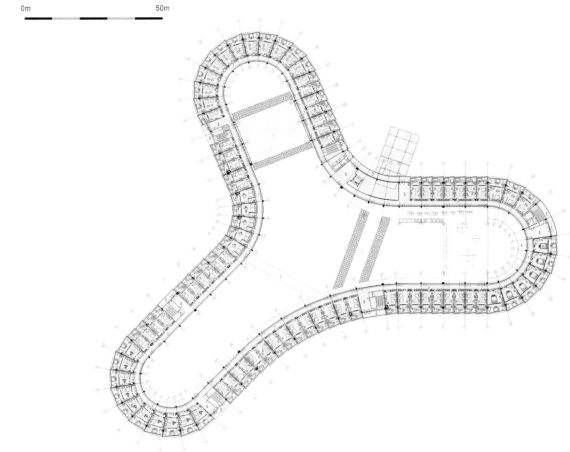

# Longmen DiPai Hot Spring Resort

龙门地派温泉酒店

设 计 师｜曾卓中、袁润铣、葛新亮、冯宇彦
设计公司｜广州市品祺装饰设计工程有限公司

DiPai Hot Spring Resort is oriented to create the international music hot spring resort, so the concept of music and natural health go throughout the whole space, especially the landscape-centered designing thought. The site is large enough so there is no need of air conditioning in many spaces, such as lobby, aisles, and the hollow part of corridors, in these place the natural ventilation is sufficient.

The application of the natural landscape is the successful part of the architectural design of DiPai Hot Spring Resort. From the rill under the bridge at the entrance to the lotus pond in the lobby, and to the waterfall in the hot spring area, the scenery goes through back and forth, and the waters and sky merge in one color, integrating the indoor artificial environment with the outdoor natural environment.

Although the hotel is not large, with only 400 rooms, it is a unique hot spring resort.

With perfect grasp of the design, it is comparatively stable and coordinating on the whole, without the exaggerating design of ordinary hotels and the striking terror in your hearts. The colors and materials of rooms and corridors feature a strong local village atmosphere, highlighting the popular look and modern artistic characteristics without losing local features. The functions planning of the building is complete, where the public facilities, building facilities, services facilities, buildings and transportation hubs are linked reasonably.

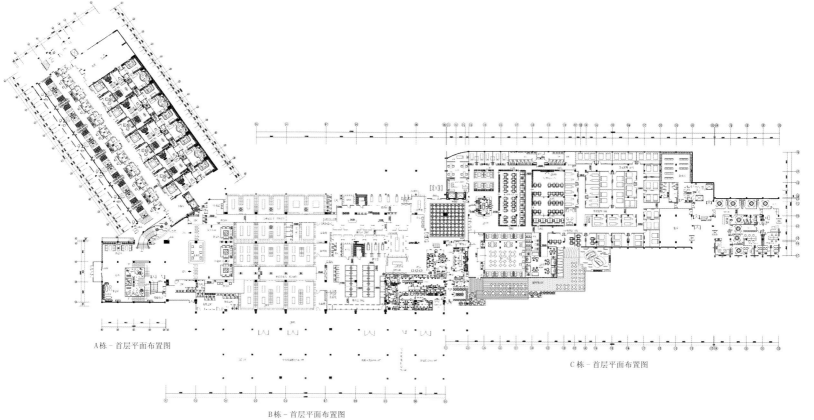

A栋 - 首层平面布置图

B栋 - 首层平面布置图

C栋 - 首层平面布置图

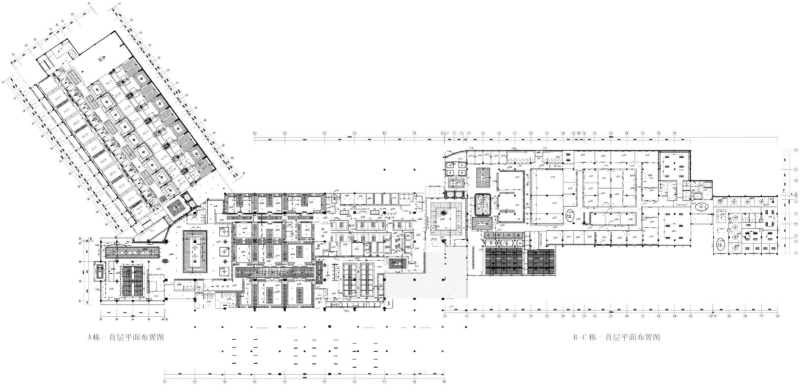

A栋-首层平面布置图　　　　　　　　　　　B-C栋-首层平面布置图

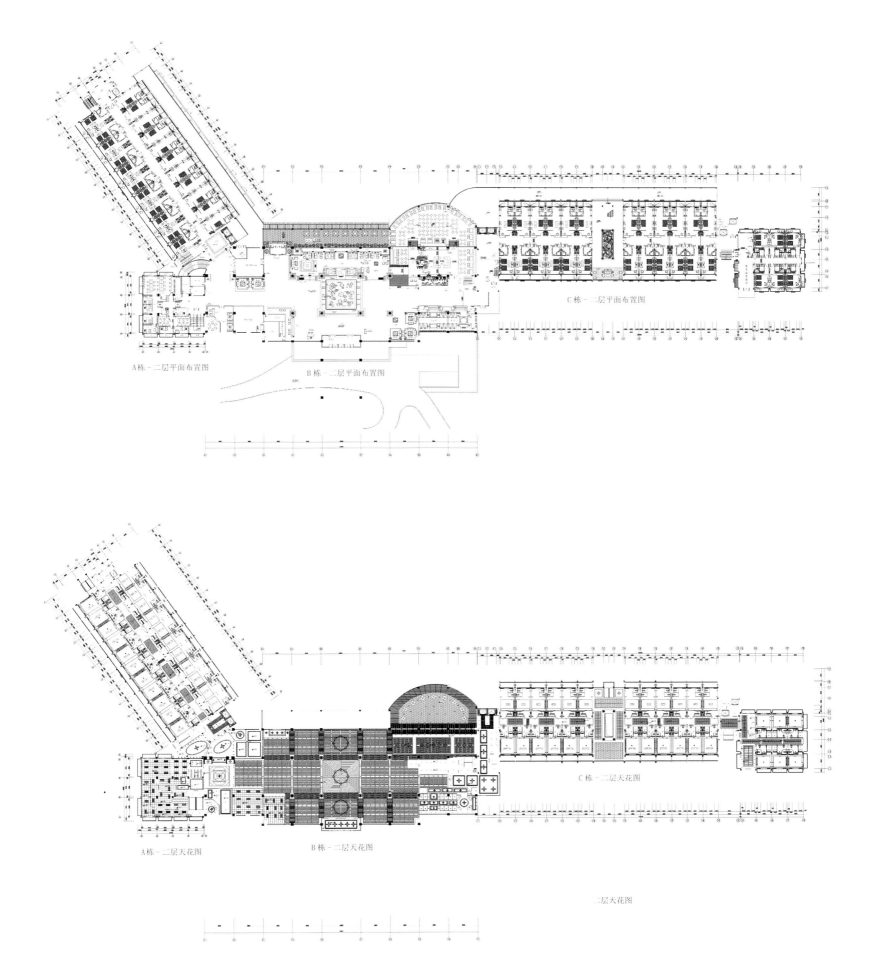

地派温泉酒店的设计定位是国际音乐度假温泉酒店，因而从一开始就贯穿了音乐、自然养生的理念，特别是以山水为中心的设计思想。建筑空间开阔，许多空间可以不用空调，如大堂、走廊、连廊中空部分，仅仅自然通风就足够。

自然景观的应用是地派温泉酒店建筑设计中一个很成功的部分。从入口处桥梁下的小溪到大堂的荷花水池再到温泉区的瀑布流水，景色的气场前后贯通、水天一色，使室内的人工环境与室外的自然环境融为一体。

酒店虽然规模不大，只有 400 多间客房，但它是一间十分有特色的温泉酒店。

分寸把握准确，并且从总体来看，比较稳重、协调，没有一般酒店设计的张扬和先声夺人。房间和走廊在色彩和材料上有强烈的当地乡村气息，既突出了酒店的流行面貌和现代艺术特征，又不失地方特色。建筑的功能规划完整、公共设施、建筑设施、服务设施、建筑和交通枢纽衔接合理。

# Wuxi Lingshan Yuanyi Ri-Star Spa Hotel

无锡灵山元一丽星温泉度假酒店

**项目地点** | 江苏无锡
**项目面积** | 45000 平方米
**设 计 师** | 姚胜虎、郭又新、朱寿耀、叶作源
**设计公司** | 上海胜异设计顾问有限公司
**主要材料** | 米黄酸洗面、片岩文化石、仿古砖、原木、自然面青石
**摄　　影** | 周跃东

The project is located in Taihu National Tourist Resort on Maji Mountain Peninsula in the southwest of Wuxi. The hot spring project is facing water with hills on the back, near to rippling Taihu Lake on the east, with the famous Lingshan scenic area on the north. The construction area of the indoor hot spring club is 15,000 square meters, and the outdoor area is divided into supporting functional area and open-air hot spring pool area.

At the beginning of the preparation of the project, the designers participated in the project's commercial positioning, and functional planning, architecture, landscape and other related designs. Under the premise of respecting nature, the designers try to build a hot spring self-cultivation tourist resort with the theme of leisure flavor. The interior design of the hot spring club is based on the strong and pure Southeast Asian style, coupled with the unique local culture in Taihu, Lingshan, and the simple and natural design technique is introduced to show the natural beauty of Zen within the space. Here, you can feel the natural beauty of Lingshan, Zen mood of Lingshan Buddha and the magnificent Taihu Lake, therefore, this is a resort for purification of mind and self cultivation.

无锡灵山三期工程配套接待中心
Wuxi Lingshan Project

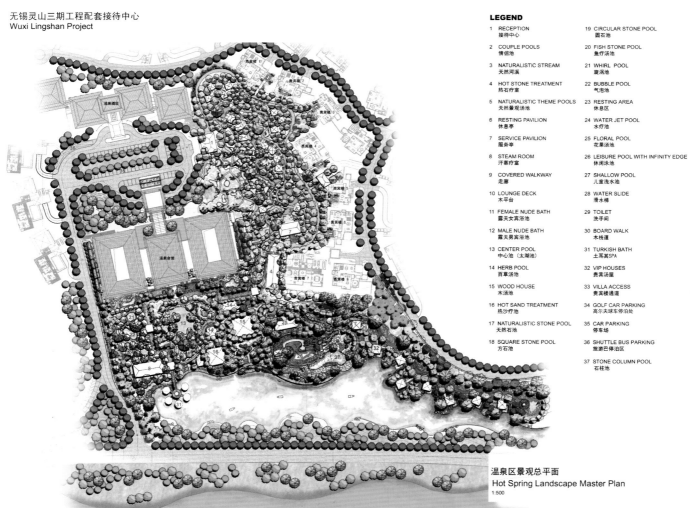

温泉区景观总平面
Hot Spring Landscape Master Plan
1:500

**LEGEND**

1. RECEPTION 接待中心
2. COUPLE POOLS 情侣池
3. NATURALISTIC STREAM 天然河滨
4. HOT STONE TREATMENT 热石疗室
5. NATURALISTIC THEME POOLS 天然景观汤池
6. RESTING PAVILION 休息亭
7. SERVICE PAVILION 服务亭
8. STEAM ROOM 汗蒸疗室
9. COVERED WALKWAY 走廊
10. LOUNGE DECK 木平台
11. FEMALE NUDE BATH 露天女宾浴池
12. MALE NUDE BATH 露天男宾浴池
13. CENTER POOL 中心池（太湖池）
14. HERB POOL 百草汤池
15. WOOD HOUSE 木汤池
16. HOT SAND TREATMENT 热沙疗池
17. NATURALISTIC STONE POOL 天然石池
18. SQUARE STONE POOL 方石池
19. CIRCULAR STONE POOL 圆石池
20. FISH STONE POOL 鱼疗汤池
21. WHIRL POOL 旋涡池
22. BUBBLE POOL 气泡池
23. RESTING AREA 休息区
24. WATER JET POOL 水疗池
25. FLORAL POOL 花果汤池
26. LEISURE POOL WITH INFINITY EDGE 休闲泳池
27. SHALLOW POOL 儿童浅水池
28. WATER SLIDE 滑水梯
29. TOILET 洗手间
30. BOARD WALK 木栈道
31. TURKISH BATH 土耳其SPA
32. VIP HOUSES 贵宾汤屋
33. VILLA ACCESS 贵宾楼通道
34. GOLF CAR PARKING 高尔夫球车停泊处
35. CAR PARKING 停车场
36. SHUTTLE BUS PARKING 旅游巴士停泊区
37. STONE COLUMN POOL 石柱池

　　该案位于无锡市西南端马迹山半岛的太湖国家旅游度假区内。温泉项目背山面水、东临千顷碧波的太湖、北面是著名的灵山圣境风景区。室内温泉会馆建筑面积15000平方米，户外配套了功能区和露天温泉汤池区。

　　设计师在项目的筹备之初就参与了项目的商业定位以及功能规划、建筑、景观等相关专业的设计配合。项目在尊重自然的前提下，致力打造以度假休闲风情为主题的温泉养身旅游胜地。温泉会馆的室内设计以浓郁纯正的东南亚风格为蓝本，结合灵山太湖特有的地域文化，用朴实自然的设计手法表现出空间的自然禅意之美。在这里，人们可以感受灵山自然之秀美、灵山大佛之禅境、太湖之瑰丽，因此，这里是净化心灵、修身养性的度假圣地。

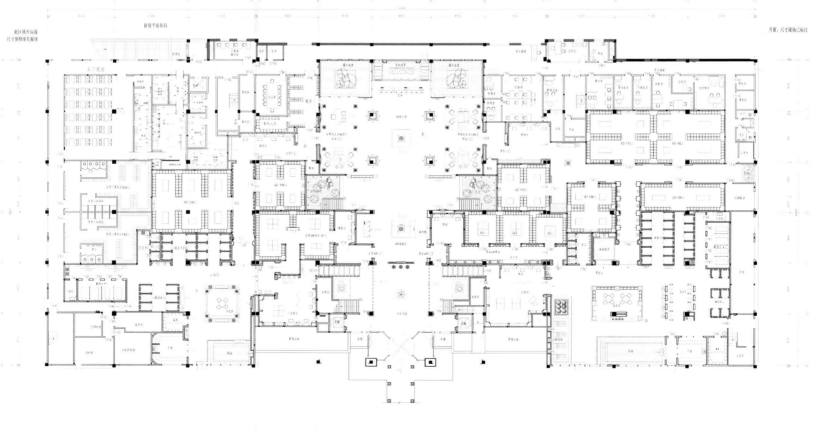

一层平面图

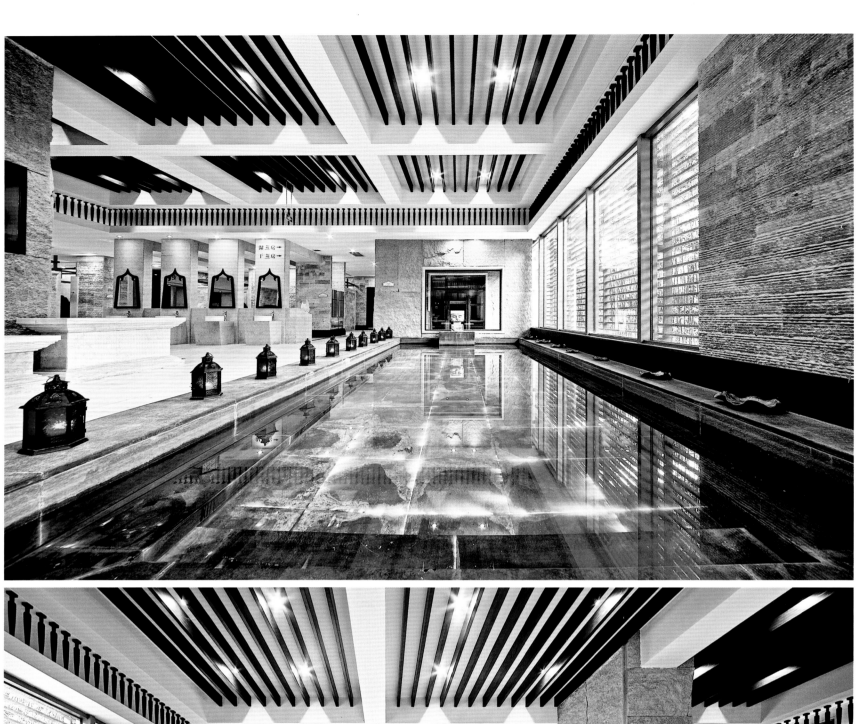

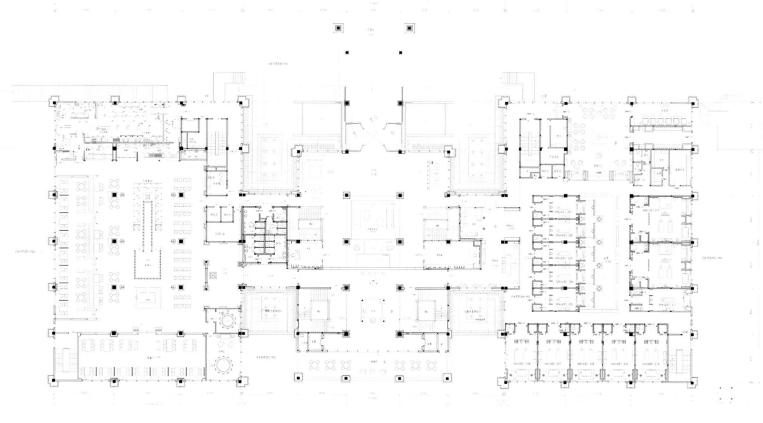

二层平面图

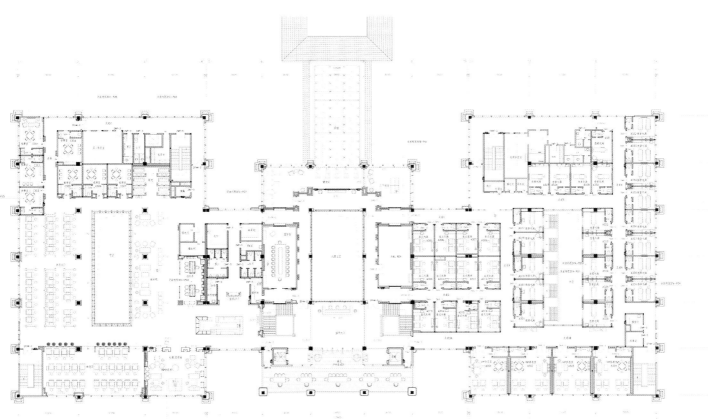

三层平面图

# Mineral Garden and Spa

矿泉花园温泉

设 计 师 | Mario Botta
客　　户 | Credit Suisse Anlagstiftung, Zurich
用　　户 | Aqua-Spa-Resorts, Development & management AG, Bern
项目经理 | MLG Generalunternehmung AG, Bern
摄　　影 | Enrico Cano

Rigi Kaltbad presents a natural landscape of extraordinary beauty. The mountain overlooking the Lake of the Four Forest Cantons commands a beautiful view dominated by the peaks of the Central Alps.

This original environment permitted to house a project in close relation both with the natural context and the village of Rigi Kaltbad.

The project consists of two interrelated parts – the village square and the thermal baths and a spa beneath the square. The thermal bath will be accessible for everybody; the spa for adults only.

The new village square appears as a "mineral garden" embellished by eight skylights that bring the natural light in the interior, and by a vast observation deck from which the visitors can enjoy a beautiful view.

The square is delimited to the east by the planned railway Vitznau-Rigi and to the west by a circular tower that represents both the access to the facility and the link to the cableway to Weggis.

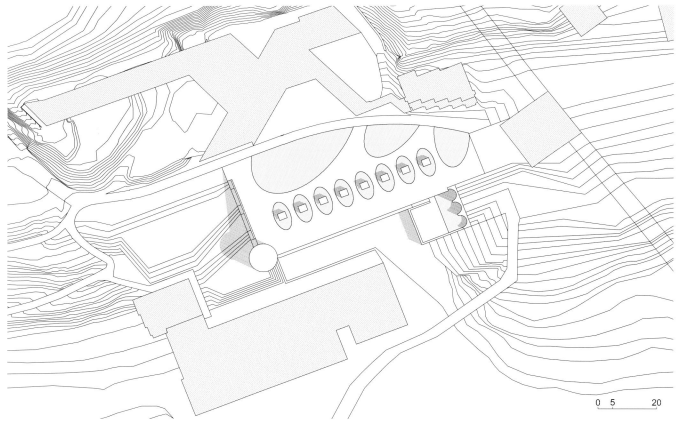

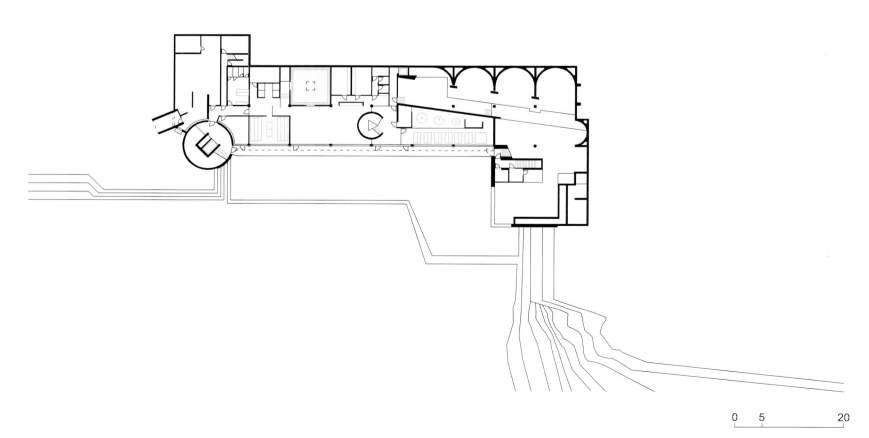

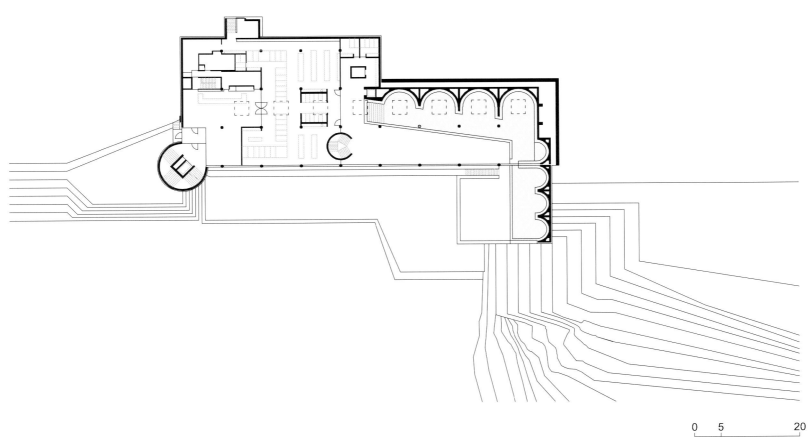

0  5  20

瑞吉·卡尔特巴德展现了自然景观的非凡美丽。从山上可俯瞰四州湖，欣赏阿尔卑斯山峰的美丽景色。

设计师利用原始环境修建该项目，使其与自然环境及瑞吉·卡尔特巴德村有着密切的关系。

该项目包括两个相互关联的部分——村庄广场和温泉浴场及广场下面的水疗中心。温泉浴场可供每位顾客使用；而水疗中心只对成人开放。

新的村庄广场以"矿泉花园"的形象出现在人们面前，饰有八个天窗，使自然光线照射到室内。顾客可以在巨大的观景台上欣赏美丽的风景。

计划修建的菲茨瑙·瑞吉铁路将广场与东部区分开来，而与西部之间则隔着一座代表着设施通道及韦吉斯索道链接处的圆形塔。

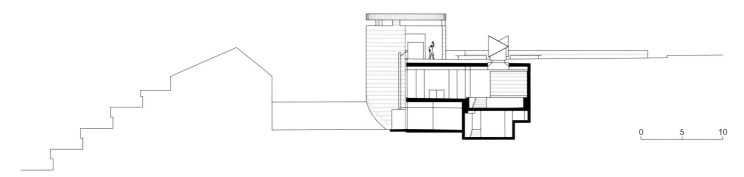
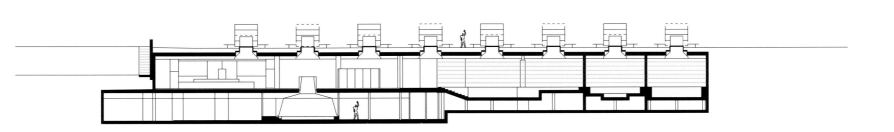

# Dino-valley Hot Spring Resort Changzhou

常州恐龙谷温泉酒店

项目地点 | 江苏常州
项目面积 | 15579 平方米
建筑方案设计 | 洲联集团（WWW5A）·五合国际
室内设计 | 深圳市寅界建筑室内设计有限公司
开发商 | 龙城旅游控股集团有限公司

Dino-valley Hotspring Resort Changzhou takes hot spring as the core, integrating SPA, dining, meeting, living and other functions as a whole. In planning, with the layout principle of "landscape around the building, building around courtyard", the green belt encloses the hot spring reception center, dining center, spa club and other groups of buildings. At the same time, the group of public buildings and the spa and living area behind enclose the outdoor hot spring park, so the buildings and the landscape are interpenetrated, existing side by side and playing a part together. The public buildings along the city roads are designed with metal and glass, and respectively mean the hot spring rocks and water, shaping a creative architectural style and reflecting the designing purpose of "ecology, nature, fashion and elegance".

The interior design continues the architectural theme to build the whole space with canyon morphology, and on the basis of full use of natural elements, it introduces simple and fashionable design.

On the ceiling of the lobby, LED lights are inlaid to create a starry effect, making guests feel like staying in the fantasy universe.

The restaurant is designed with a natural and casual style, where the ground and ceiling are paved with carbonized wood, eco-friendly, natural and fresh. It is surrounded by glass embellishment of ferns patterns, with multi-layer composite lighting, modern and warm.

The rest area and SPA are themed with leaves and wood, introducing natural materials coupled with multi-layered lighting to achieve natural and warm space effect. The natural sisal and logs give guests a sense of returning to nature.

On the left of the lobby is the natural and modern guestroom area. It features suites and standard rooms, and some rooms are designed with supporting courtyard and introduce pool into the room. The Nordic-style wooden frame background and the warm-light droplights on both sides of the bed make the whole atmosphere warmer. The rooms are equipped with mini bar, wardrobe, desk and other supporting facilities, making guests indulged in pleasure.

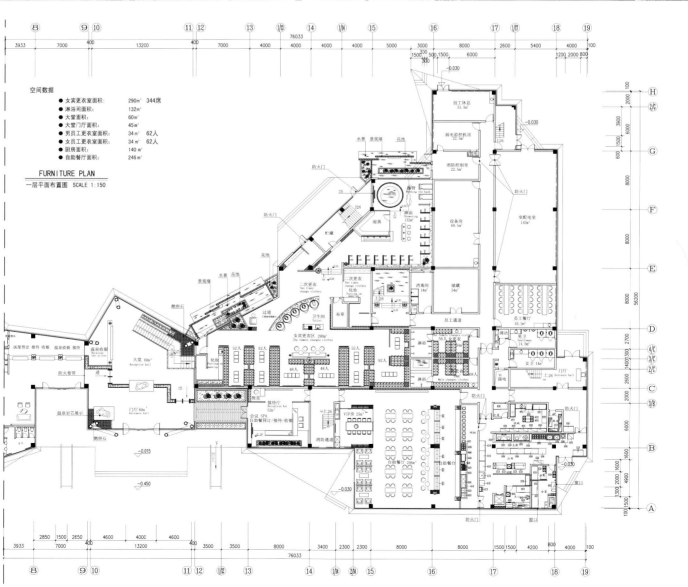

空间数据
- 女宾更衣室面积: 290m² 344席
- 淋浴间面积: 132m²
- 大堂面积: 60m²
- 大堂门厅面积: 45m²
- 男员工更衣室面积: 34m² 62人
- 女员工更衣室面积: 34m² 62人
- 厨房面积: 140m²
- 自助餐厅面积: 246m²

FURNITURE PLAN
一层平面布置图 SCALE 1:150

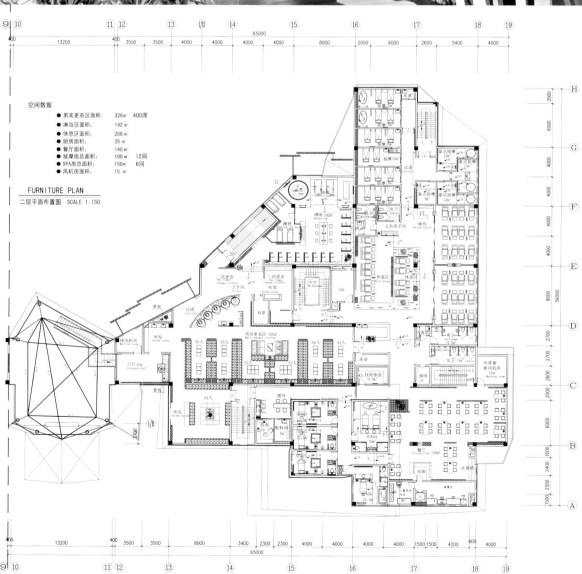

空间数据
- 男宾更衣区面积: 326 m²  400席
- 淋浴区面积: 142 m²
- 休息区面积: 206 m²
- 厨房面积: 35 m²
- 餐厅面积: 140 m²
- 按摩房总面积: 100 m²  12间
- SPA房总面积: 150 m²  6间
- 风机房面积: 15 m²

FURNITURE PLAN
二层平面布置图  SCALE 1:150

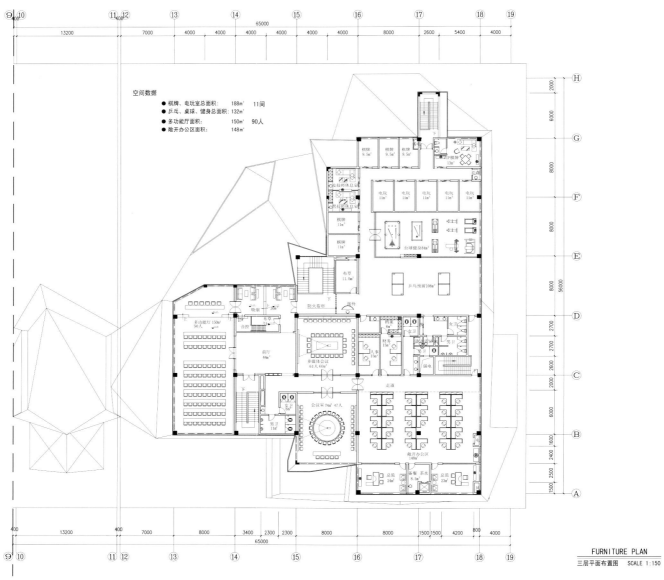

FURNITURE PLAN
三层平面布置图　SCALE 1:150

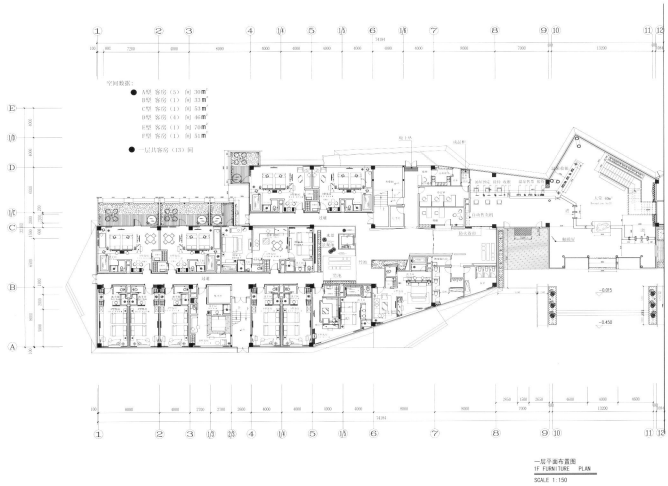

一层平面布置图
1F FURNITURE PLAN

SCALE 1:150

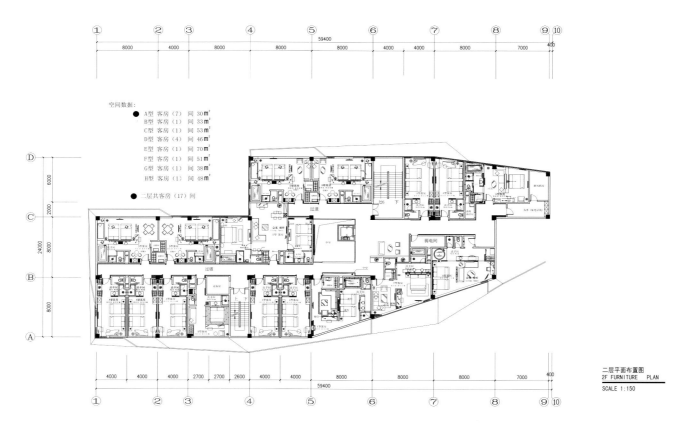

空间数据：
- A型 客房（7）间 30㎡
- B型 客房（1）间 33㎡
- C型 客房（1）间 53㎡
- D型 客房（4）间 46㎡
- E型 客房（1）间 70㎡
- F型 客房（1）间 51㎡
- G型 客房（1）间 38㎡
- H型 客房（1）间 48㎡
- 二层共客房（17）间

二层平面布置图
2F FURNITURE PLAN
SCALE 1:150

三层平面布置图
3F FURNITURE PLAN
SCALE 1:150

常州恐龙谷温泉酒店以温泉为核心，集水疗、餐饮、会议、住宿等功能为一体。规划上以"景观围绕建筑，建筑围合庭院"的布局原则，绿化带围绕温泉接待中心、餐饮中心和温泉会所等几组建筑。同时这组公建和后面的汤宿区包围着室外温泉公园，使得建筑与景观相互渗透、相辅相成。沿城市道路的公共建筑以金属和玻璃分别寓意温泉岩石和水，塑造出一个富有创意的建筑造型，体现了"生态、自然、时尚、典雅"的创作主旨。

室内设计延续建筑主题，以峡谷形态构造整个空间，在充分借用自然元素的基础上，融入简约时尚的设计。

大堂天花板上，LED灯缀饰出繁星效果，令人恍如置身于梦幻的宇宙。

餐厅采用自然休闲风格，地面天花由碳化木地板铺成，既生态环保又自然清新。四周用蕨类植物纹样玻璃点缀，多层复合灯光照明，既现代又温暖。

休息区、水疗区以叶、木为主题，采用自然材质配以多层次灯光照明以达到自然温馨的空间效果。天然的剑麻和原木，给人回归自然的感觉。

大堂左侧是自然现代的客房区。房型设计分套房和标准房，在一些房间中配套庭院，并将泡池引入室内。北欧风格的木架背景、床头两侧暖色光的吊灯，使整体气氛更显温馨。房内配置了迷你吧、衣柜、写字台等配套设施，令人流连忘返。

# Fujian Strait Culture Homeland Hot Spring Resort

## 福建海峡文化原乡温泉度假村

**项目地点** | 福建福州
**项目面积** | 333333 平方米
**设 计 师** | 郭文河
**设计公司** | 广州市文和园林景观设计有限公司
**主要材料** | 植物用材：棕榈、鱼骨葵、三药槟榔、老人葵；花区：厚皮香、桂花、羊蹄甲、木棉、红千层等；果区：龙眼、荔枝、芦柑、橄榄
硬质用材：现场的河石、鹅软石、旧枕木、天然实木、火山石、石板

The designer pays attention to the sense of order in density and distribution, to make the whole space full of vitality. At the same time, the approach of combining functional space and landscape space is introduced to meet the functional use and achieve the effect of shaping characteristic landscape. The environment design in partial space takes full consideration on humanities requirements and viewpoint elements. The occasion relation of garden landscaping is used to create mutual penetration and connection, to achieve the effect of different scenes on different steps. It mainly contains flowers shelf gallery, sprinkler sculpture and other parts, creating a refined and pleasant exchanging space. In the water landscape, the fresh water in pools is shimmering, adding another charm with careful savour.

Different functions are clearly and reasonably distributed by groups of users taking consideration of the needs of different functions without disturbing each other, to highlight the function and practicality. Various kinds of plants in orderly density are interspersed in the site, and the distinctive scene wall with running water and sculptures are used to embellish the place. The changes of flooring form in different materials are organically, ecologically, naturally and harmoniously integrated the project into the surrounding environment. The hot spring is built along the creek, and taking into account that people need a rest space when enjoying the scenery and people's hydrophilic nature, the designer places pools of different functions on both sides of the creek.

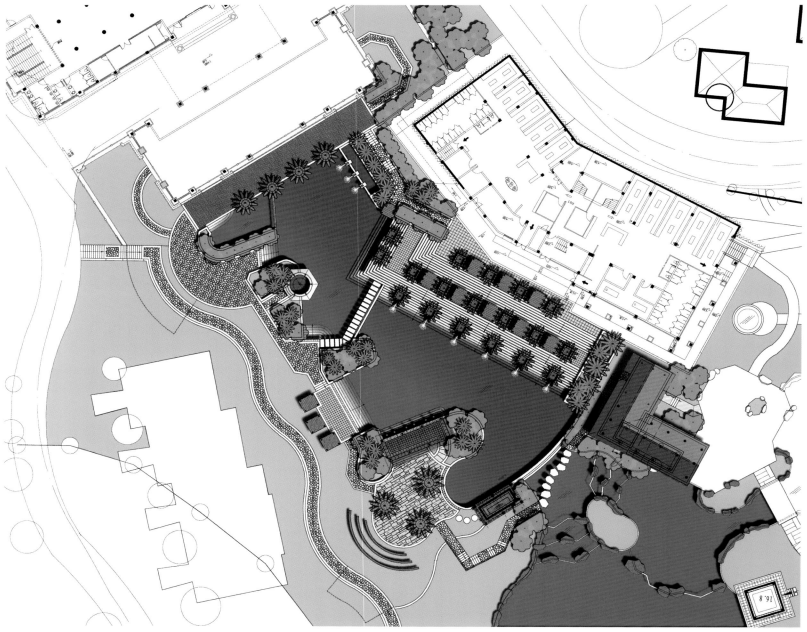

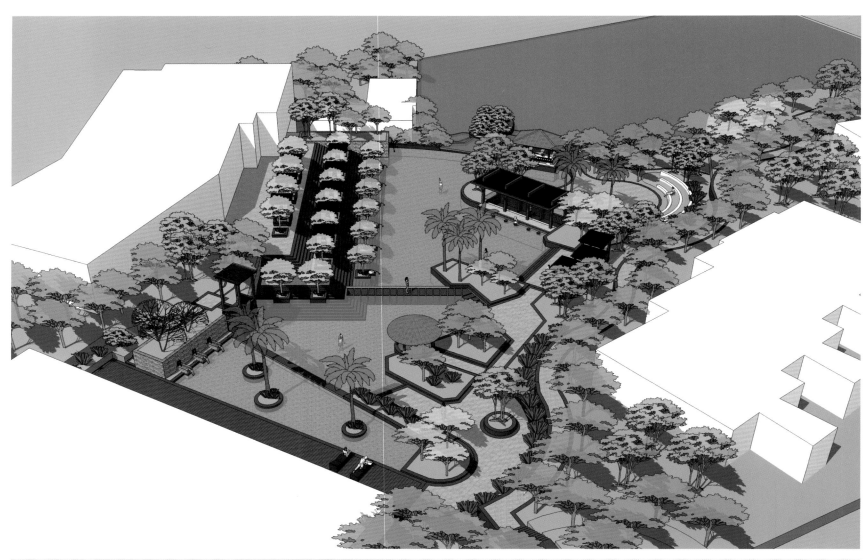

　　本案注重空间疏密、集散的秩序感，使整个空间充满活力。同时采取了功能空间和景观空间相结合的手法，既满足了功能使用性，又做到了塑造特色景观的效果。在局部空间环境设计上，充分考虑人文要求和视点因素。利用园林造景的应景关系，进行相互渗透和衔接，达到一步一景、步移景异的效果。主要包括花架廊、喷水雕塑等几部分，营造了精致而宜人的交往空间。水景区灵动的池水波光闪烁，再次细细品味，更添一番意趣。

　　各区功能明确，按使用人群合理分布，照顾到不同功能的需要，相互之间互不干扰，突出了功用性与实用性。再通过品种丰富、疏密有序的植物穿插种植，独具匠心的流水墙、雕塑小品的安排点缀，材质各异的铺地形式的衔接变化，有机、生态、自然、和谐地将其溶入到周边大环境之中。温泉临溪而建，考虑到游客们在赏景时需要一个休息的空间以及人们的亲水性，还在溪涧两旁设计了不同功能的温泉。

# Global Villa Hot Spring Paradise

锦绣香江温泉城

项 目 地 点 | 广东广州
项 目 面 积 | 总面积：300000 平方米；温泉面积：25000 平方米
规划及建筑设计 | 广州山橡设计顾问有限公司
温泉及景观设计 | 广州市山橡景观设计工程有限公司

The project is located in the scenic spots area of Zengcheng Whitewater Stockade, which is known as the magnificent emerald at the Tropic of Cancer. This place features year-round prosperous trees, and the anion content in the air reaches as high as $141,000/cm^3$, the highest one in southern China, and it is known as the "green lung in Pearl River Delta". The hot spring paradise collects the lovely views in Nankun Mountain National Forest Park, enjoys the highest waterfall in mainland China, and embrace the delicate beauty of Mount Wofo. "Thousands of mountains are standing and hundreds of springs are spewing, with eight rivers around the stockade and a waterfall in the area," the unique natural scenery gives birth to the culmination of the perfect vacation place in Global Villa Hot Spring Paradise.

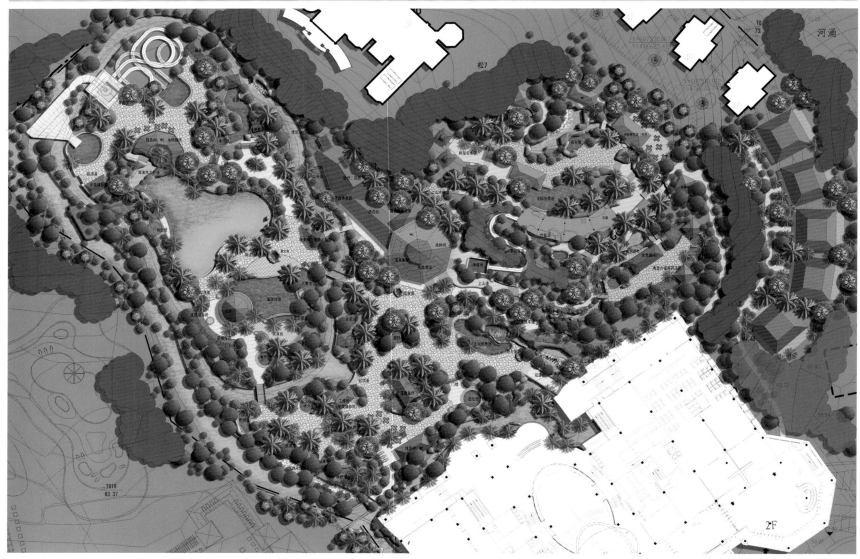

该案坐落在"北回归线上的瑰丽翡翠"——增城白水寨风景名胜区内。区域内林木终年繁盛，空气中负离子含量高达 14.1 万个 /$cm^3$，为华南地区之最，素有"珠三角绿肺"之称。温泉城尽收南昆山国家森林公园秀丽美景，拥揽中国大陆第一高瀑，感悟卧佛山灵秀之气。"千峰林立，百泉喷涌，八水绕城，一瀑环拥"，独特的自然风光，孕育了锦绣香江温泉城极尽完美的度假佳境。

# Zhongshan Hot Spring Hotel

中山温泉宾馆园林改造

项 目 地 点 | 广东中山
项 目 面 积 | 总面积：110000 平方米；温泉池面积：20000 平方米
温泉及景观设计 | 广州市山橡景观设计工程有限公司

**Planning pattern: one axis, two different landscapes, and multiple features**

One axis refers to the cultural landscape axis designed in the front square of the hotel lobby, namely the entrance waterscape, lawn plaza, and great men's footprint which are introduced throughout the front square of the lobby. The design changes the original fuzzy layout of the hotel, highlighting the main reception area and making the traffic flowing line clearer.

Two different landscapes: two different landscapes refer to the layout where the inner landscape and the outside environment complement each other. Luo Sanmei Mountain outside the hotel is the first distant view, one of the natural conditions in the whole park; the existing lake is used to re-build landscapes through piling slopes and leaving a corridor for introducing landscape, thus forming the second view; the two embrace and complement each other, reproducing the view borrowing in Lingnan Garden.

Multiple features: for the various building types in the park, the combinations of spaces are not the same, and the hotel is designed according to circumstances and shows the characteristics of each space, creating different features in different places.

### Creation of Lingnan Garden Style

1. The aim of the garden is to re-build a hot spring resort with historical and cultural deposits, and characteristics of new Lingnan;

2. In condition of meeting the functions of the hotel, it retains the relations and space structure of the original building;

3. It reserves the features of pillars structural frame, whole-decorated facades and decorations of Lingnan-style building;

4. It continues and develops the characteristics of Lingnan garden;

5. The main materials of the garden are blue-gray, echoing with the contextual background;

6. Planting, reserving the skeleton of the old trees to highlight the mood of Lingnan, and taking advantage of the local tree species to ensure the color and smell of the seasons.

### Use of New Ideas and New Technologies

The hotel takes advantages of geothermal energy of hot spring and recycles the thermal energy of water resources, linking with the air conditioning heating and forming the complement of energy. In the design, the overflowing water from hot spring and the water recycling pool are connected, and the thermal energy of recycling water supports the hot kang to save energy.

**规划格局：一条轴线，两重山水，多园特色**

一条轴线，是指酒店大堂前广场空间，设计了一条文化景观轴线：入口水景、草坪广场、伟人足迹的元素，贯穿整个大堂前广场空间。这一设计，改变了原来酒店空间布局模糊的旧貌，突出了酒店的接待主体，使交通流线更加清晰。

两重山水：两重山水是指园内园林与外部环境的相互借景格局。酒店外的罗三妹山是远景的第一重山水，作为整个园区的自然条件之一；利用现有的湖区，重新堆坡造景，并留出借景的视线走廊，这是第二重山水；两者相互映衬相互辉映，实现岭南园林的借景。

多园特色：因为园区的建筑类型多样，形成的组合空间也不一样，设计因地制宜，发挥各自空间的特色，营造出不同园区的不同特色。

**岭南园林风格的营造**

1. 园林的宗旨是，重建一个具有历史文化沉淀的、具有新岭南特色的温泉度假酒店；

2. 在满足酒店功能使用的条件下，保留原建筑的群落关系和空间结构；

3. 保留岭南派建筑的梁柱结构、整饰立面和装修特点；

4. 延续发展岭南园林的风格特点；

5. 主要园林材料以青灰色为主，呼应文脉背景；

6. 绿化种植、保留古树的骨架、突出岭南的意境，利用本土树种优势，保证四季的色、香、味。

**新理念新技术的运用**

酒店利用温泉地热的优势，把水资源的热能进行回收利用，与酒店空间的空调取暖相联动，形成能量的互补。园林设计中还让温泉的溢水与废水回收池相连，回收废水的热能提供温泉的热炕等利用，节省能源。

# Tianmu Lake Tuanbo Holiday Hotel

## 天沐湖团泊假日酒店

**项目地点** | 天津
**项目面积** | 总面积：2000000 平方米；温泉酒店：78000 平方米
**设计公司** | 北京维拓时代建筑设计公司

The whole project covers a total area of about 2,000,000 square meters, where the hot spring hotel covers 78,000 square meters, and the construction land is located in the modern agricultural demonstration garden of the Tuanbo New City touring town, Jinghai, Tianjing, adjacent to Duliujian River, with 2,400 meters away from Tuanbo Lake, featuring rich water resources.

The design introduces "implicating" culture as the theme, starting from the cultural concept of "harmony between man and nature, returning to nature", and designing around the context of "island house with hidden springs", to create a green, humanizing and high-quality resort hotel.

Along the rear side of the artificial islands, the groups of buildings are arranged in order and separated from each other. Facing the main entrance at the central is the largest functional area – hot spring hotel area, with hotel lobby, meeting, dining and other public facilities and 160 five-star standard guestrooms of various types. On the right side of the hotel is the hot spring area, and the left side is the VIP club area, providing bathing, dining, and accommodating services to high-end customers. Under the VIP club area is the enterprises club area.

The outbound traffic flow of the artificial island is mainly vehicle stream, with stepwise shunt way. Each partition features a relatively independent parking place, meeting the parking needs of various functional areas nearby. The logistics freight corridor is hidden in the landscape, connecting with the underground loading area through a ramp, to ensure the integrity of the landscape on the ground.

On the island, the flow of people is placed along the buildings near the lake, achieving the effect of different scenes in different steps when passing the footpath, micro relief, and trestle bridge.

Adopting the Tang style, this program introduces traditional culture of the Chinese health hot spring, and its well-proportioned bodies are naturally integrated with the environment. The architectural courtyard and the space composing way complement with the Tang style, highlighting the "implicating" culture – the theme of Chinese culture.

Stone, metal decorative components and paintings are collocated with each other, and the fresh and natural tone shows the exquisite and light modern materials in the traditional charm, achieving the effect of showing the traditional culture spirit through modern building materials.

　　整个项目总占地约 2000000 平方米，其中温泉酒店用地 78000 平方米，建设用地位于天津静海县团泊新城旅游特色镇组团中的现代农业示范园，毗邻独流减河，距团泊湖 2400 平方米，水利资源丰富。

　　设计以"隐"文化为主题，从"天人合一，回归自然"的文化理念出发，围绕"岛院隐泉"的脉络进行设计，打造绿色、人文、高品质的休闲度假酒店。

　　建筑群沿人工岛后侧依次布置并各自独立分区。中央正对主入口的是体量最大的功能区——温泉酒店区，设有酒店大堂、会议、餐饮等公共配套设施和 160 间五星级标准的各类客房。温泉酒店区右侧为温泉区，温泉酒店左侧为 VIP 会所区，为高端客户提供洗浴、餐饮、住宿服务，VIP 会所区下方为企业会所区。

　　人工岛的对外交通流线以车流为主，采用逐级分流的方式。每个分区均有相对独立的停车场，就近满足各个功能区的停车需求。后勤货运通道隐藏于景观之中，与地下室卸货区通过坡道联系，保证地上景观的完整。

　　岛上人行流线沿建筑临湖面布置，通过步道、微地形、栈桥等达到步移景异的效果。

　　建筑风格取法中国温泉养生的传统文化，采用唐式风格，其错落有致的形体与环境自然交融，建筑院落和空间的构成手法也与唐式风格相得益彰，凸显"隐"文化这一中国文化的主题。

　　建筑采用石材、金属装饰构件和涂料搭配形式，色调清新自然，于传统韵味中体现现代材料的精致感和轻盈感，达到用现代建筑材料体现传统文化精神的效果。

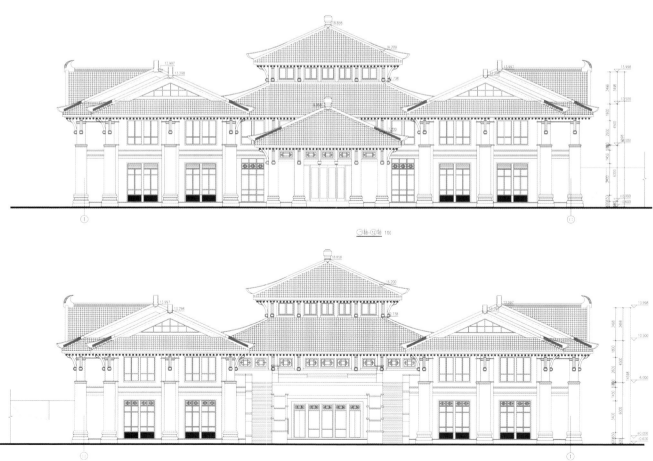

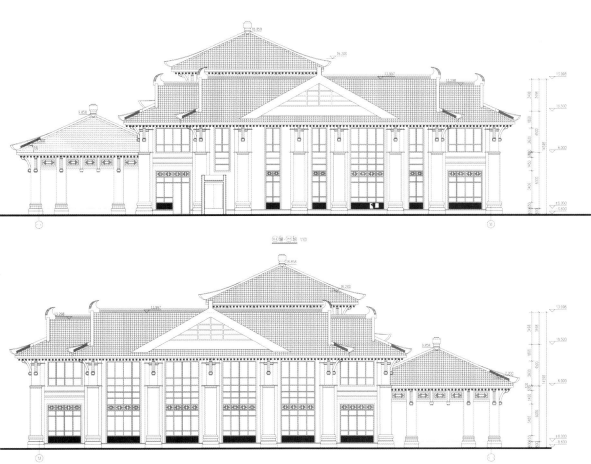

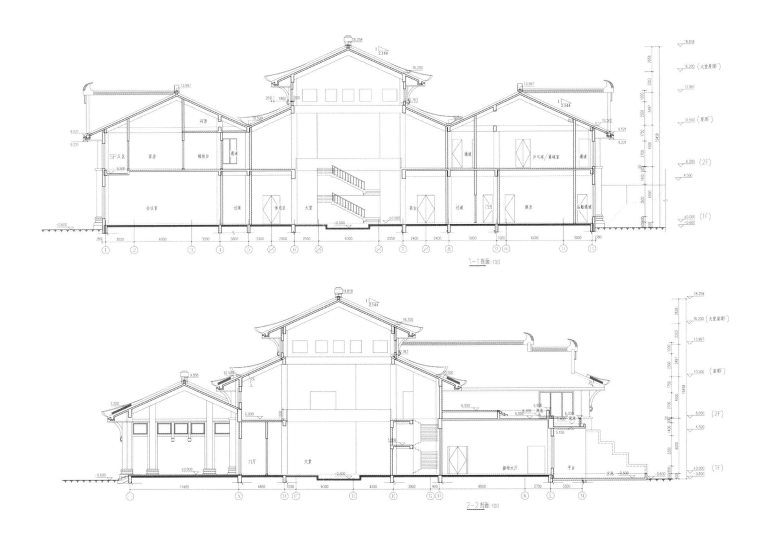

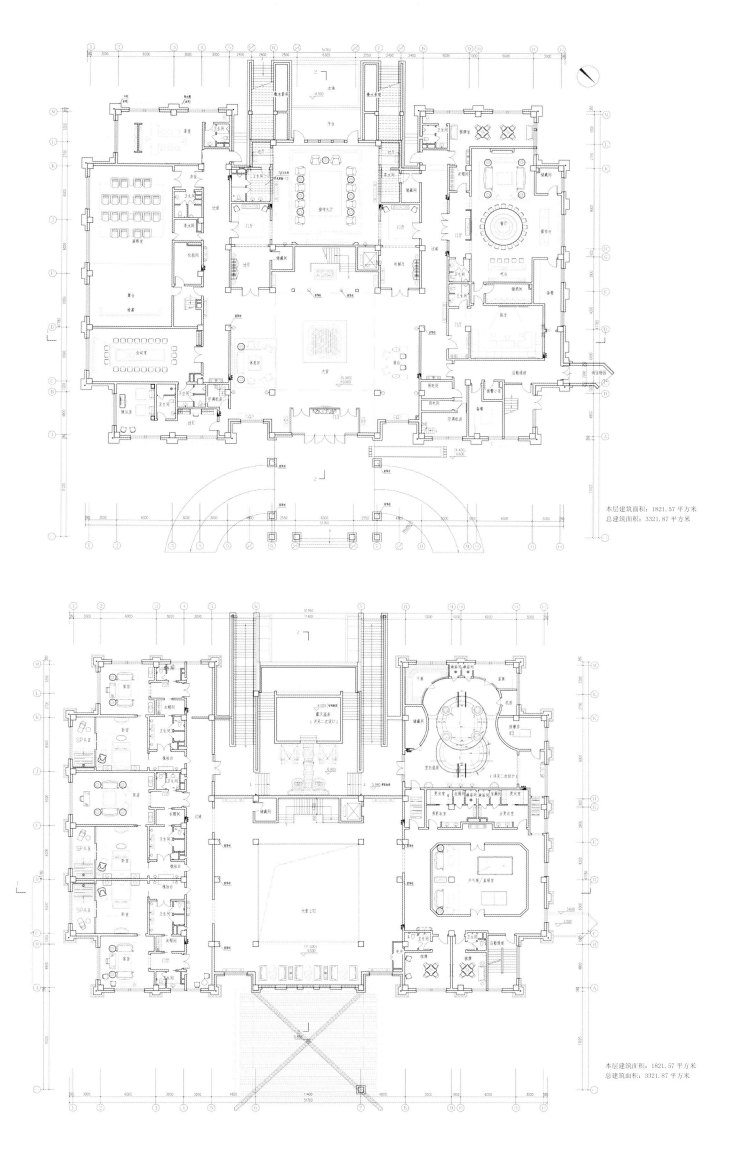

# Tianyi Hot Spring Resort

天一温泉度假村

**项目地点** | 福建连城
**项目面积** | 3333333平方米
**景观设计** | 厦门艺道景观规划设计有限公司
**开 发 商** | 福建省冠豸山旅游发展有限公司

Tianyi Hot Spring Resort is a natural sloping hot spring integrated with traditional Hakka humanities, outlining the unique quiet, elegance and harmony of hot spring through humanistic and natural symbols. It is built in rolling hilly basin, just as the old saying goes, "swaying lotus and green grass are linked with beads, and a bunch of hot springs are built on a bunch of happy lands". In the fog, you can see the flickering crowded Hakka architectures, ancient material finishes, small decorations of natural style, and the windmill and the waterfront filled with living atmosphere. With the luxuriant green and murmuring of running water, you can shuttle in the quiet and mystical roads and bath in fragrant pools, just like staying in the land of idyllic beauty, relaxing without any thinking or worry.

The design scale of Tianyi Hot Spring Resort is large, and the mountain and water landscape patterns are built along the lake. According to the terrain, the moving lines and functional partition are reasonably planned, and the characteristics of Hakka culture and local customs are combined to introduce the Hakka theme throughout the design with distinct characteristics.

　　天一温泉，一座融入客家传统人文的自然坡地温泉，用人文和自然的符号勾画出温泉独有的静雅与和谐。天一温泉在起伏的丘陵盆地之间，正应了"风荷翠盖联珠秀，一串福地一串汤"这句话。迷雾之中，忽隐忽现的是鳞次栉比的客家建筑、古朴的材料饰面、自然风格的小品，还有充满生活气息的风车水岸。绿意葱葱，水声潺潺，穿梭在幽密的曲径中，沐浴在清香的汤池里，恍如世外境域，神情倏然，无思无虑。

　　天一温泉园区设计规模较大，沿湖建成山水相映的景观格局。依据地形合理设置动线与功能分区，并结合客家文化特征及当地的人居习俗，使客家的主题贯穿始终，特色分明。

# One & Only Hotel

开普敦 One & Only 度假酒店

项 目 地 点 | Waterfront, Cape Town, South Africa
设　　计　师 | Fabian Architects
室内设计师（卧室与公共区域）| Tihany Design
项 目 经 理 | SIP Project Managers
酒店开发商 | Kerzner International

The hotel concept is that of an "urban resort" consisting of a ninety-two suite tower on the urban edge and two islands within the Marina encompassing a further forty Villa style suites, a spectacular landscaped swimming pool and spa complex.

The uniquely landscaped Spa Island separates the Villa Island with its pool and private restaurant from the tower. The main hotel (referred to as "the tower") houses the Presidential and Imperial suites (both in excess of 250 sqm.) and is crowned at the two uppermost floors with three magnificent duplex Penthouses. Located at the entrance level are the exciting Lobby Bar, the all-day dining Rubens and Speciality Nobu Restaurants. The tower is positioned to complete the string of highly sought after apartment buildings that surround the marina.

The building design therefore results in a single loaded corridor at each level giving access to the guest rooms. The plan shape of the tower exaggerates the gentle semi crescent curve of the existing marina thereby exposing the Porte Cochere as it peels away to enhance the visitor's sense of arrival.

Under the direction of Kerzner International with its intense attention to detail, a collaboration of more than thirty consultant firms many of which are from other parts of the globe have produced an urban resort hotel in Cape Town which truly deserves the title of One & Only.

# LAYOUT PLAN

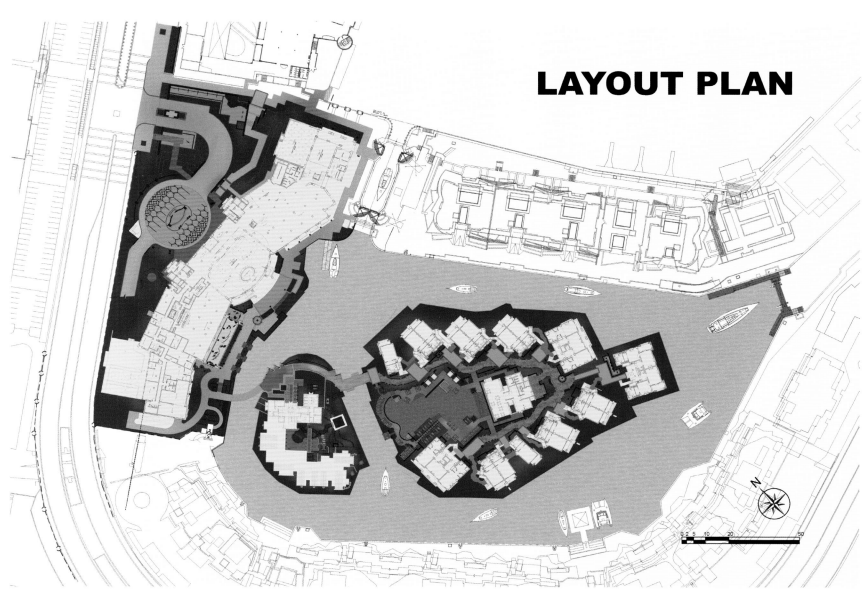

**NORTH ELEVATION**

**SOUTH ELEVATION**

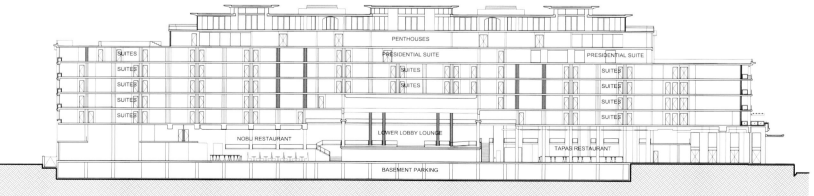

**SECTION EE**

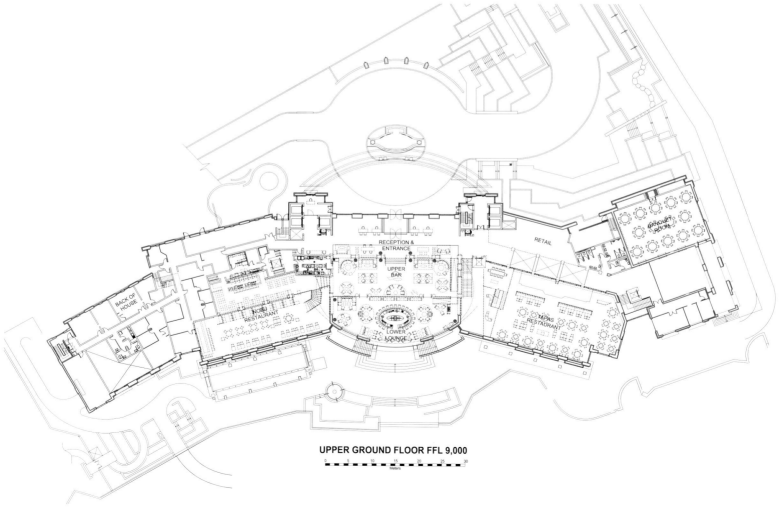

**UPPER GROUND FLOOR FFL 9,000**

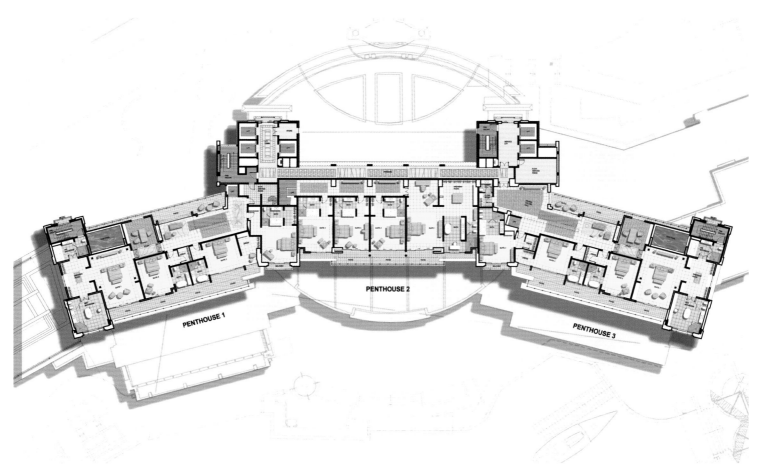

**LEVEL 6 - PENTHOUSE LAYOUTS**

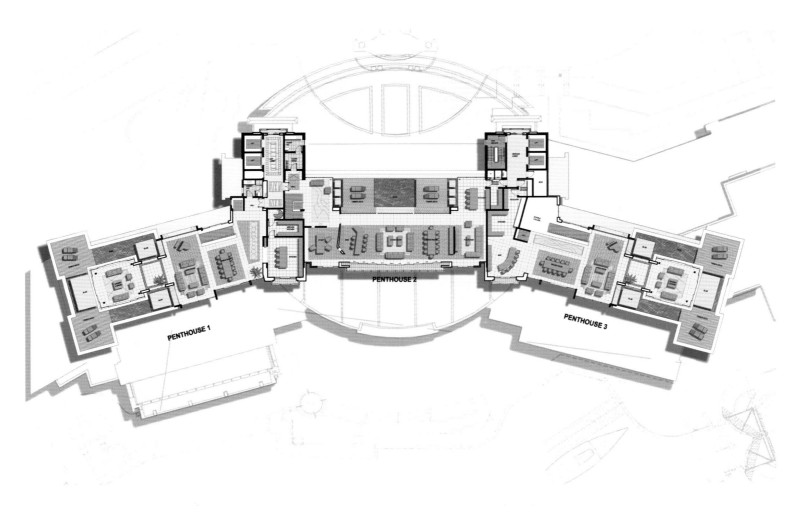

**LEVEL 7 - PENTHOUSE LAYOUTS**

以"城市度假胜地"为理念，该酒店坐落于城市边缘，拥有92间塔楼套房，在码头的2个岛屿上还设计了40间别墅式套房、一个壮观的景观游泳池和一组水疗中心。

独特的园林温泉设有游泳池和私人餐厅，将别墅岛与套房区分开。酒店主区（以下简称为"塔楼"）设有总统套房和帝王套房（面积均超过250平方米），并且在最上面的两层楼内还设计了三间奢华复式公寓。坐落于入口层的是让人兴奋不已的大堂酒吧、鲁本斯24小时全营业餐厅和专业的Nobu餐厅。套房区旨在完善环绕码头的系列紧凑式公寓楼。

因此，该建筑设计最终在每层都提供了单独的走廊，通往客房。塔形设计夸大了现有码头那大方的半月形曲线，凸显了供车辆出入的门廊，也加强了游客宾至如归的感觉。

Kerzner国际非常注重细节，在其指引下，来自全球各地的30多家顾问公司在开普敦建立了这个城市度假胜地——名副其实的One & Only度假酒店。

# Hot Spring Leisure City

温都水城

**设计公司** | 澳大利亚 SDG 设计集团
**客　户** | 宏福集团

Hot Spring Leisure City is located in original site of Pingxi Mansion, on the north central axis of Beijing, on the north of it is the only natural river flowing through Beijing—Wenyu River, and to the southeast and southwest, two to three kilometers away are Tiantongyuan and Huilongguan, two of the national largest cultural living zone. The place features rich hot spring resources, laying foundation for the success of the project.

### Water Space

Water space is the large-scale water theme park in Hot Spring Leisure City. With its enormous rectangular shape, its curve ceiling and modern architectural style, it stands out from the surrounding groups of classical architectures. Water Space targets the public, families, children and youngsters as the main consumers. With a construction area of approximately twenty thousand square meters, it can accommodate up to four thousand people to entertain themselves inside the building. So far in Asia, it is the most advanced hot spring, aquatic recreational amenities with the theme of water.

### Hot Spring Health Club

The building area is approximately 8,000 square meters, with simple Chinese style. The whole building has two floors, expanding around the central outdoor hot spring courtyard, where the first floor of wet area is distributed into fitness area, hitting area, swimming area, and bathing area, with thirty five functional bathing pools in all, and it creates a natural effect of tropical forest and rocks cave; the second floor of dry area is divided into places of hot spring body beauty and hot spring face beauty, hot spring healthy nutrition restaurant, aromatherapy, jade bed, resting hall and so on. It is connected with the water space and the commercial street through a corridor.

### Outdoor Hot Spring Area

It is divided into three separate hot spring areas with construction area of about 2,510 square meters. The corridor with antique flavor connects the three outdoor hot spring pools together, and the surrounding quadrangle courtyards features 44 guestrooms separately. Each room is designed with an independent outdoor hot spring pool, and such pattern creates a private, cozy and comfortable leisure space for guests.

### Hot Spring Quadrangle Courtyard

The nine sets of quadrangle courtyards are designed with Japanese style and Chinese style separately, with a total building area of 4,790 square meters, including five sets of Chinese style courtyards and four sets of Japanese style, and the building area of each set is about 530 square meters. Each set is designed with independent indoor and outdoor hot spring pools and sauna rooms, and the courtyard features high-grade small meeting room, family-style self-catering kitchen, standard room and suite. There are blooms of flowers, trees and bamboo, stepping pebbles, providing an elegant and unique landscape. Each courtyard can accommodate six to twelve guests, this is a land of idyllic beauty for business meetings, family travel vacation and friends gathering.

The outdoor hot spring area and hot spring quadrangle courtyards are surrounded by moat, so guests can go into the space only through the special drawbridges, which ensures the privacy of the guest. The three special docks along the moat allow guests to travel around the whole water town by boat.

"温都水城"地处平西王府旧址，在北京的北中轴线上，北面是唯一流经北京的天然河流——温榆河，东南、西南两三公里便是天通苑、回龙观这两个全国最大的文化居住区。这里还具有丰富的温泉资源，让该项目具备了成功的基础。

## 水空间

水空间是温都水城中的一个大型水主题公园，它巨大的矩形体量、曲线形的屋顶和现代建筑风格在周围古典建筑群体中十分醒目。水空间以大众、家庭、儿童和青少年作为主要的消费对象，其建筑面积约2万平方米，可以同时容纳近4000人在馆内进行娱乐消费，是亚洲目前最先进的以水为主题的温泉水上娱乐场馆。

## 温泉养生会馆

建筑面积约8000平方米，为简约中式风格。整个建筑分两层，围绕中心室外温泉庭院展开：一层湿区按健身区、击打区、泳区、泡区等共设置了35个功能性泡池，并营造了热带丛林和山石溶洞的自然效果；二层干区设有温泉美体和温泉美容、温泉养生营养餐厅、香薰、玉石床、休息大厅等。它通过连廊与水空间和商业街相连。

## 室外温泉区

共分三个相对独立的温泉区，建筑面积约2510平方米。古色古香的连廊将三个室外露天温泉泡池区串联起来，环绕周边的四合院每套分别有44个房间，每间都带有独立的室外温泉泡池，私密性的格局为客人营造出惬意舒适的休闲空间。

## 温泉四合院

9套四合院分别采用日式和中式风格，总建筑面积约4790平方米，其中中式5套，日式4套，每套建筑面积约530平方米，均具有独立的室内外温泉泡池和桑拿浴房，院内配有高档小型会议室、家庭式自助厨房及标准间、套间。小院内花树竹丛、卵石汀步、景观优雅别致。每套四合院内可容纳6～12人住宿，是商务会议、家庭旅游度假和朋友聚会的世外桃源。

室外温泉区、温泉四合院被护城河水系所环绕，只有通过专用的吊桥才能进入该区域，保证了客人的私密性。沿河的三个专用码头可使客人乘舟游览整个水城。

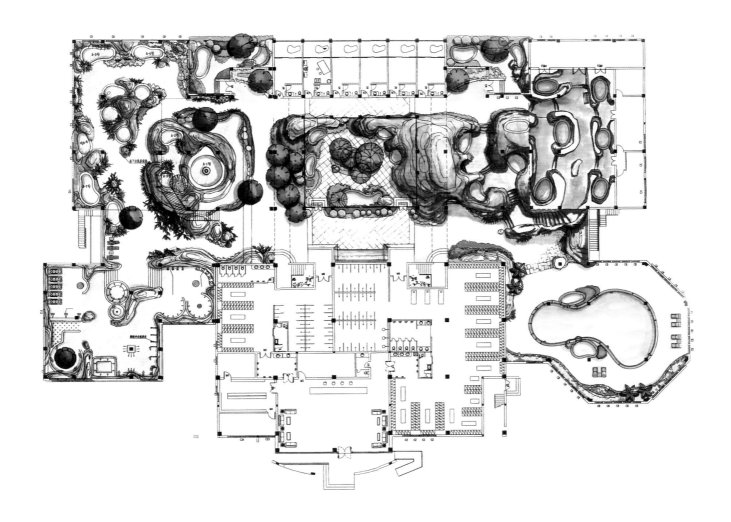

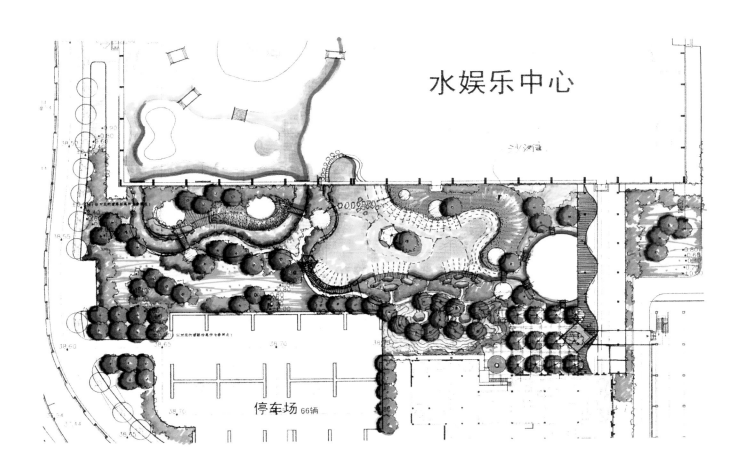

水娱乐中心

停车场 66辆

# Wuxi Lingshan Yuanyi International Resort

无锡灵山元一国际度假村

项目地点 | 江苏无锡
项目面积 | 300000 平方米
开 发 商 | 元一集团
景观设计 | ACLA Limited

The project contains hotel, conference center, and eco hot spring area of Southeast Asian flavor located in the southeast part of the project. The planning of the whole landscape is based on a sustainable design, optimizing the various functional spaces and strengthening green land through diversification of plants, and an artificial wetland is designed to prevent form algae spreading.

The project focuses on Doubletree by Hilton, the north and south sides of the hotel are surrounded by artificial water, to strengthen the combination of landscape and Taihu Lake water culture. The landscape for wedding studios on the north side of the hotel is composed of a series of gardens with varied characteristics, providing photo scenes changing with seasons for the guest.

In addition to the tropical features, the hot spring spa area in the project also integrates with Chinese five elements, separately introducing metal, wood, fire, water, and earth as the designing theme. In addition, the area is further divided into special hot spring area, swimming pool and children's water playing area, and in the forest there exists romantic and unique private pools for sweethearts to enjoy.

The VIP area is located in the north part of the project, and the design introduces 17 stylish arbors, and rich layers of plants to create an exquisite vacation villa garden, and to provide a private vacation space for the guest.

The whole project is composed of different parts, and each individual part is connected by the green plants and rolling water on the south side, echoing with each other. The natural and changing landscape is integrated with modern Southeast Asian vacation concept to provide a comfortable and pleasant resort for the visitor.

# 无锡灵山三期工程配套接待中心
## Wuxi Lingshan Project

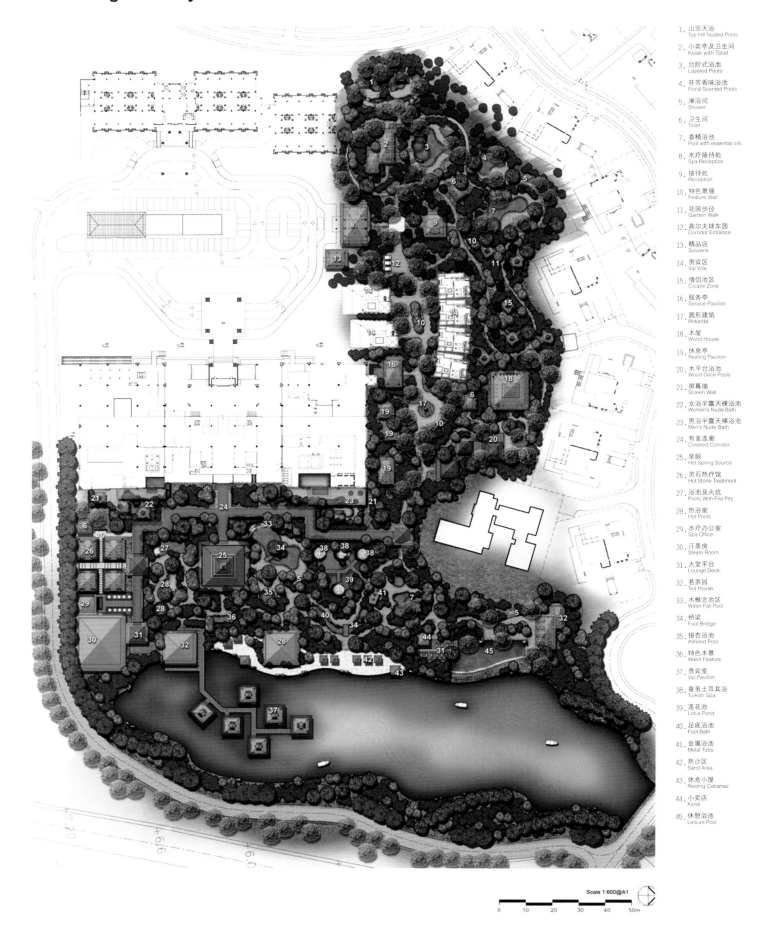

## 温泉区景观总平面
## Hot Spring Area Landscape Master Plan

无锡灵山三期工程配套接待中心
Wuxi Lingshan Project

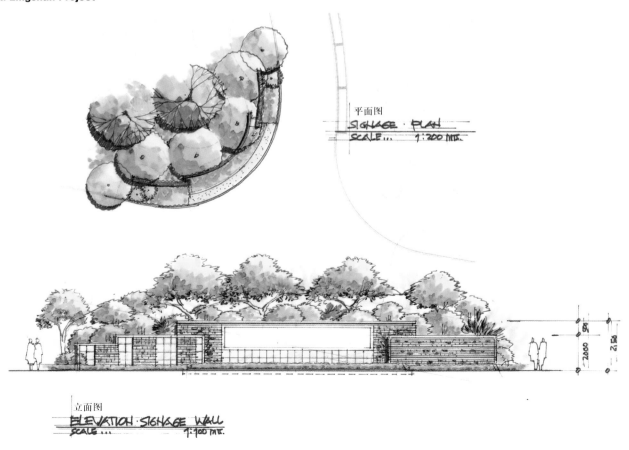

温泉区——道路入口标志墙
Hot Spring Phase 3 SPA Entrance Signage

无锡灵山三期工程配套接待中心
**Wuxi Lingshan Project**

景墙立面图
**Feature Wall Elevation**

位置图
**Location**

项目包含酒店、会议中心和东南部分富有东南亚特色的生态温泉区。整体的景观规划以可持续性为设计基础,通过多元化种植优化各个功能空间、强化绿地,设计人工湿地防止海藻蔓延。

项目的重点是希尔顿逸林酒店,人供水体环抱酒店南北两侧,加强了景观与太湖水文化的结合。酒店北侧的婚纱影城景观,由一系列丰富多变的特色花园组成,为顾客提供随四季变化的婚摄场景。

项目中的温泉水疗区除了具有热带风情特色之外,还融入中国五行元素,分别围绕金、木、火、水、土作为设计主题。另外,区域还划分为特色温泉区、游泳池和儿童水上游乐区,并在森林内设有浪漫独特的私人泡池,供情侣休憩享受。

项目北部是贵宾楼区域,设计上利用17个特色的稻草屋及丰富的植物层次打造出精致的度假别墅园林,并为住户提供私密的度假空间。

整体项目由各个不同的部分所组成,每个单项均由翠绿的植林及南侧的连绵水体连接,相互呼应。自然多变的景观融入现代东南亚的度假概念,给访客提供了一个舒适愉快的度假胜地。

# A Garda Lake Hotel with Ayurvedic Style

加尔达湖阿育吠陀风格酒店

设计师 | Alberto Apostoli
摄　影 | Maurizio Marcato

In Brenzone, on the Venetian side of Lake Garda, the Consolini Group decided to give the Belfiore Park Hotel a decidedly ayurvedic dimension by opening a new wellness centre spread along an entire floor of the hotel and named the Dhara Wellness. The project aimed to create a balance between the classical elements of the SPA by carefully studying spaces, forms, materials, colors and lights. The architect Alberto Apostoli, responsible of the project, commented: "The space is very homogeneous from an aesthetic point of view but it is also full of innovative solutions, and small details that were constructed with. I care and respect for the territory combined with a new cultural approach". By taking advantage of the local materials, the particularities of the location and a natural spring that was discovered during the construction, the SPA obtains a sensorial environment that is alive with every construction detail and material that was specially designed for it. Following a characteristic horizontal element made with bamboo canes that run along the entire floor leading to the SPA, one enters, or rather "filters in", a socializing room that is neutral and contemporary. A final element is the large olive tree that was placed within the structure during the intervention; it was originally from the land adjacent to the Hotel. The geometrical and emotional

core of the SPA is a basin of water that is constantly replenished by the natural spring whose waters are utilized almost as a lining as it flows directly onto the bare stone.

Within the basin, there is a shower that is made by a particular element that is the local stone and a special Kneipp treatment composed by single pools that are also made in the local stone. The sauna, another strong element of the centre, is created entirely in stone and is characterized by a large glass partition for the unusual transfer to the basin of water. Alongside the sauna is a vapour bath, covered by a special mosaic in resin whose parts "digitally" recreate the intertwined branches of the lemon trees. The relax area is positioned at the end of the centre and offers a magnificent view of the lake that is only a couple of meters away.

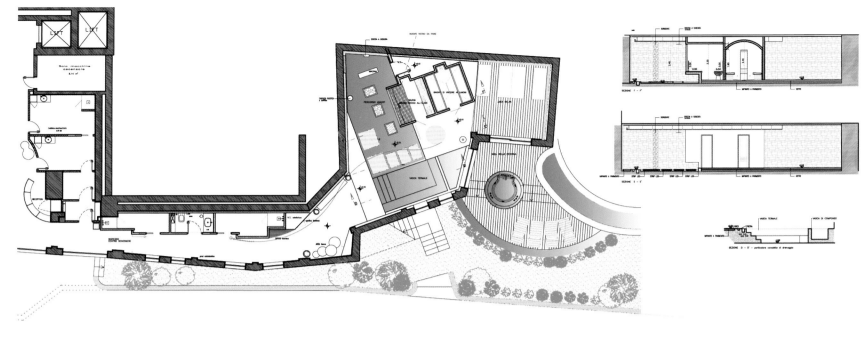

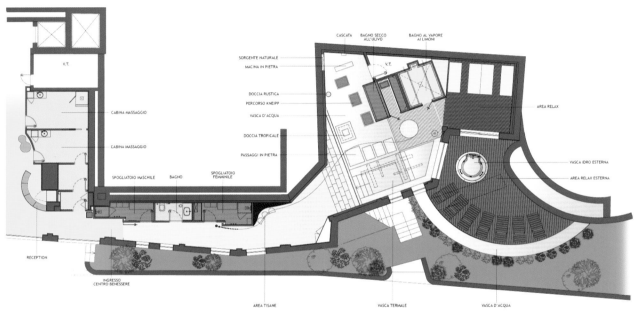

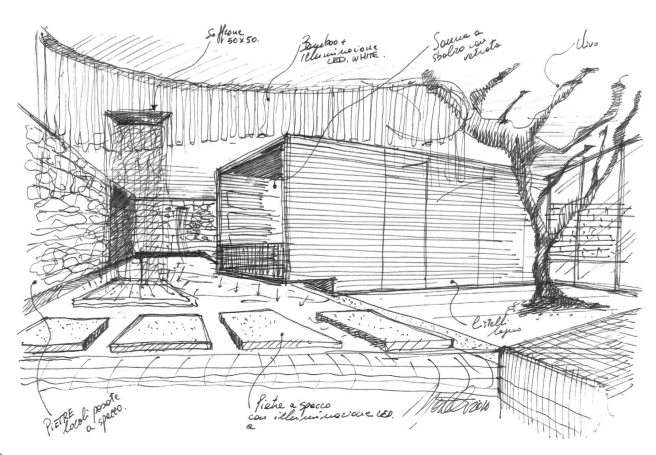

    Brenzone位于威尼斯加尔达湖畔，Consolini集团决定赋予Belfiore公园酒店以抢眼的阿育吠陀维度，设计将酒店整层楼都布置成了一个新的健身中心，并命名为达卡健身中心。该项目旨在通过认真研究空间、形式、材料、颜色和灯光之间的关系，创建出温泉经典元素之间的平衡。建筑师Alberto Apostoli负责该项目，他评论说："从美学角度来说，该空间非常均一，但它也充满了创新的设计方法，并十分注重小细节。我重视并尊重这种结合了新文化方式的设计领域。"通过利用当地材料、当地地形特色和在建筑过程中发现的温泉，该温泉最终拥有了一个很感性的环境，仿佛每一处细节建设都栩栩如生，每一种材料都是为其量身定做的一样。整层楼由别具特色的水平延伸的竹竿做成，直通温泉，带领人们进入，或者更确切地说是人们被"过滤进"中性且颇具现代色彩的社交空间。

    本案的点睛之笔是在改造过程中栽种在建筑之内的一棵大橄榄树，它是从酒店附近的土地上移植过来的。该温泉的建筑及情感核心是由天然泉水不断补充的一池水，因为泉水直接流向裸露在外的石头上，所以也被用作衬景。

    水池里有一个有特殊元素——由当地的石头构成的淋浴区，此外还有特殊的克奈普疗法，它是由许多用当地石头做成的单独汤池组成的。中心的另一个抢眼的元素是桑拿房，完全由石头构成，其最大的特点是将大幅玻璃置入水中，作为隔断使用。桑拿旁边是蒸气浴，被特殊的树脂马赛克覆盖着，并利用了数字技术再现了柠檬树枝桠交错的景象。休闲区位于中心的尽头，游客可以在此欣赏到几米之外的壮观湖景。

# Lintong · Aegean International Hot Spring Hotel

临潼·爱琴海国际温泉酒店

项目地点 | 陕西西安
项目面积 | 20000 平方米
设 计 师 | 龚小刚
设计公司 | 北京龚氏建筑设计有限公司
主要材料 | 建筑：火山浮石、红豆杉、板岩
　　　　　室内：火山浮石、西班牙米黄大理石、金丝缎大理石、雨林啡大理石、EV黑檀木饰面、玻璃砖
摄　　影 | 潘杰、李枫

The project is located in a megranate garden with 350-year history, with clubs, hot spring and villas enclosing 35 open-air pools, where the outdoor hot spring area is wholly designed with ecological and natural design concept and soothing rhythm; the indoor design features a fashionable, noble, penetrating and delicate feeling. The link between indoor and outdoor scenes is an well-regulated montage relationship, high and low, sometimes stretching, and sometimes tense.

The hot spring area consists of a combination of nine buildings, completely built by traditional craftsmen from Yunnan with the Huizhou architecture style. Each guest room is equipped with an independent outdoor hot spring pool. The nine buildings are designed with different space layout and style, allowing guests to experience for many times.

Lao Zi once said "Do nothing, but do everything". When everything is back to zero origin, is it equivalent to doing everything? The interior design is based on figure nine, to show the relationship of causal echoes between "starting point" and "finishing point", to reflect the profound Tang culture and the mood of Elysium. In the hotel lobby, the ceiling with double curved surfaces of water ripples shows the quietness of water, and the hollowed hanging volcanic pumice is taken from Tengchong, Yunnan, showing the iterative progress of how magma meets and heats underground water to form hot spring through the red cast lights behind and blue slit of light. A Tang maid is put in a niche, showing the celebrating sing and dance in Elysium. The front desk is decorated with a "Tang Quan Fu" made of hollowed bamboo, pointing to the topic.

　　该项目修建在有350年历史的石榴园内，由会所、汤泉、别墅围合着35个露天泡池，室外汤泉区全部采用生态的、自然的设计概念，节奏舒缓；室内带给人们时尚、高贵、透彻而精致的感受，室内外画面的衔接是高高低低、时而舒展、时而紧张、张弛有序的蒙太奇关系。

　　汤泉区由九座建筑组成，完全由云南传统工匠按照徽派建筑建造。每一间客房均带有独立的室外温泉泡池。九座建筑采用不同的空间布局方式和风格，让客人可以多次前往体验。

　　老子说："无为而无所不为"。当一切都回到零原点之时，是否相当于无所不为？室内设计以九为基数，表现"起点""终点"间的因果呼应关系，表现唐文化的博大精深、表现极乐世界的情境。会所大堂的双曲面水波纹吊顶表现了水的宁静、中空干挂的火山浮石取自云南的腾冲，通过背后红色的投灯、蓝色的光带表现了地下水与岩浆接触后加热升温而形成温泉的循环往复的过程。九九归一的壁龛内置唐侍女，表现了极乐世界的歌舞升平。前台用镂空竹简呈现的一首"汤泉赋"点明了主题。

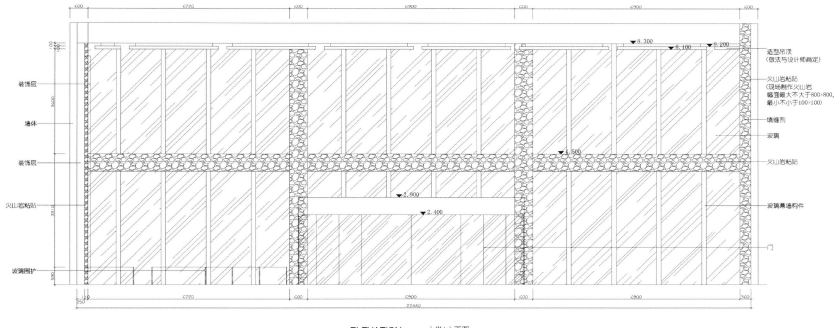

ELEVATION 大堂A立面图
SCALE 1:50

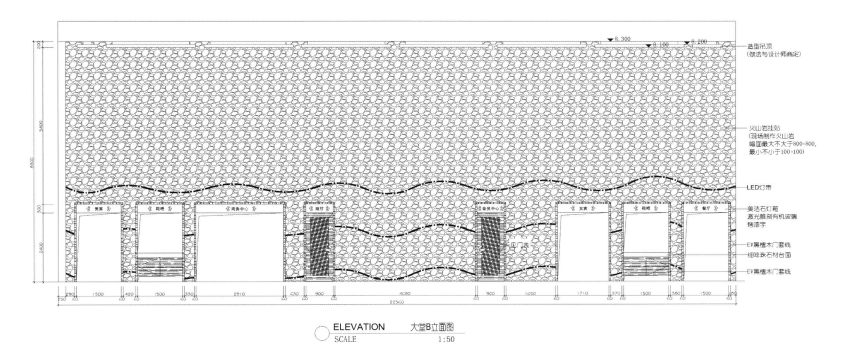

ELEVATION 大堂B立面图
SCALE 1:50

ELEVATION 大堂B立面龙骨布置图
SCALE 1:50

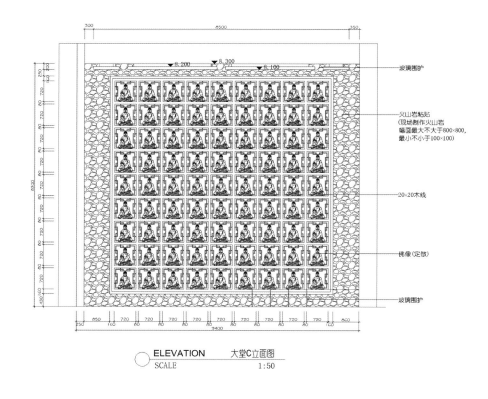

ELEVATION  大堂C立面图
SCALE  1:50

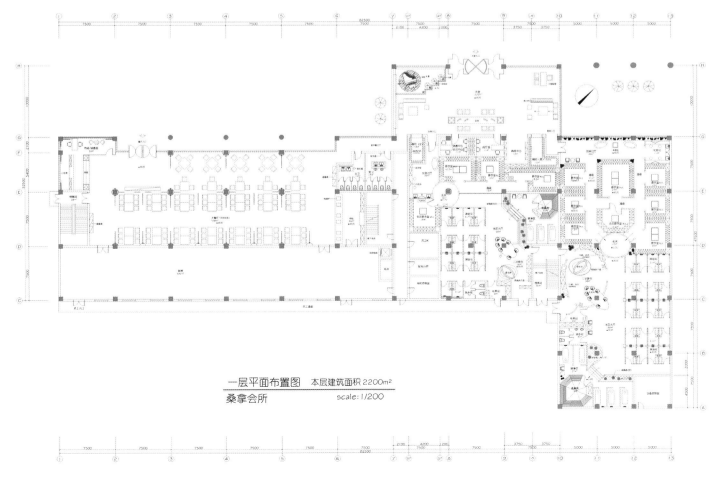

一层平面布置图　本层建筑面积 2200m²
桑拿会所　　　　scale:1/200

# Nanjing Yuhao Tangshan Hot Spring International Hotel

南京御豪汤山温泉国际酒店

项目地点 | 江苏南京
项目面积 | 29000 平方米
设 计 师 | 孙彦清、肖莹、胡坤、毛邦凯
设计公司 | 苏州金螳螂建筑装饰股份有限公司
主要材料 | 大理石、木饰面、墙纸、皮革、地毯
开 发 商 | 南京城市建设开发集团

The project is a hot spring resort decorated with the architectural style of the 1920s and 1930s, which is built according to a five-star standard, and it is located in Tangshan, Nanjing, the best one of the four hot spring capitals in China. The hotel is divided into south building and north building, where the south building features five floors on the ground, six floors in some parts, and one floor underground; the north building features four floors on the ground and two floors underground.

The unique style of 1920s and 1930s shows the hotel's magnificent overall shape, and highlights the unique, distinguished and elegant hotel.

Into the lobby, a three-storey atrium comes into your eyes, while the beige marble, fine copper pillars with grillwork, and the landscape in the atrium complement each other, making the hotel clean, simple, elegant and warm, giving you a feeling of a home away from home. The hotel has various kinds of 164 elegant and comfortable rooms, in addition to the special executive floor and hot spring suites, another three sets of luxurious villas with different styles of decorating are the first choice to entertain the guests and honor dignitaries. In the hotel, the Chinese and Western restaurants, banquet private rooms, specialty restaurant, café, high-grade bar and other dining venues provide all kinds of delicacies for the guest. The multi-functional rooms and various kinds of meeting rooms are equipped with advanced audio-visual equipments, to meet the reception needs of different sizes and different types. The hot spring bathing center, gym, tennis courts, mini golf, snooker room, KTV, chess room are the ideal choices to relax body and mind and enjoy a healthy and leisurely life. The various kinds of open-air hot spring pools are scattered in the shade of trees and embedded in the roof gardens, bringing guests many surprises about skin care and health.

Lead a royal life and enjoy the luxury. The hotel respects its guests and nobles, human feeling of cheerful landscape will endow the guests at home and abroad with fresh enjoyment!

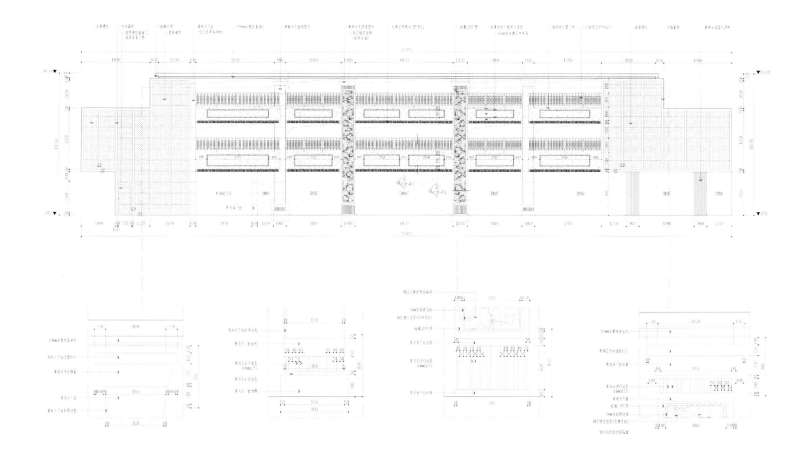
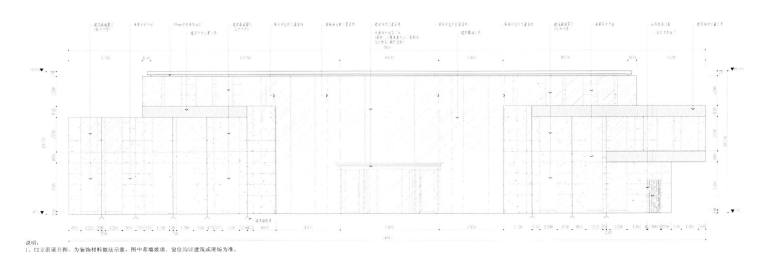

说明:
1、E2立面展开图,为饰面材料做法示意,图中幕墙玻璃、窗位均以建筑或现场为准。

E2 ELEVATION

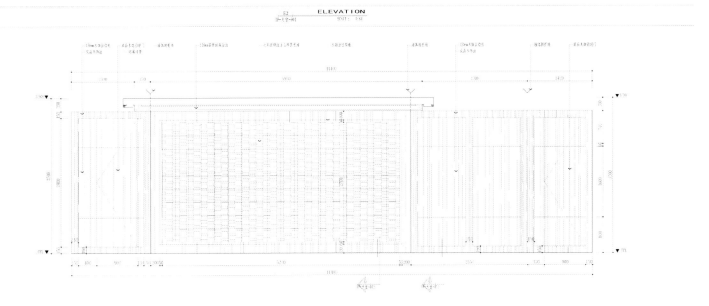

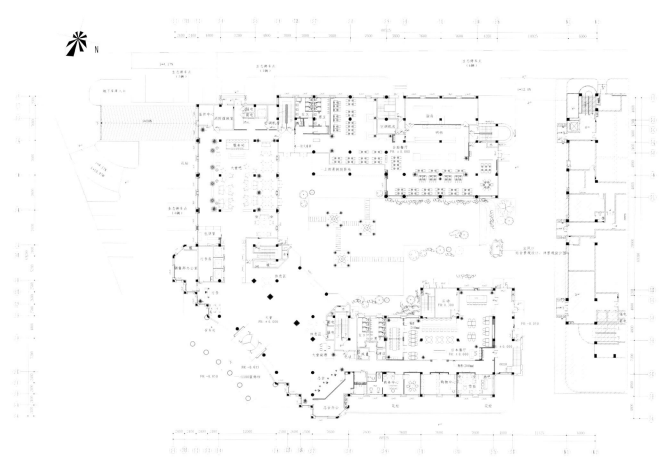

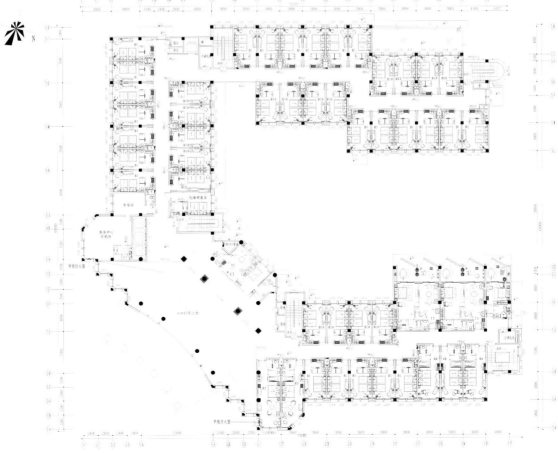

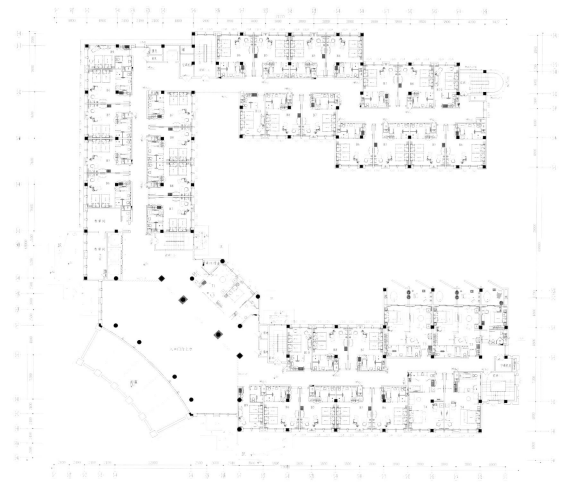

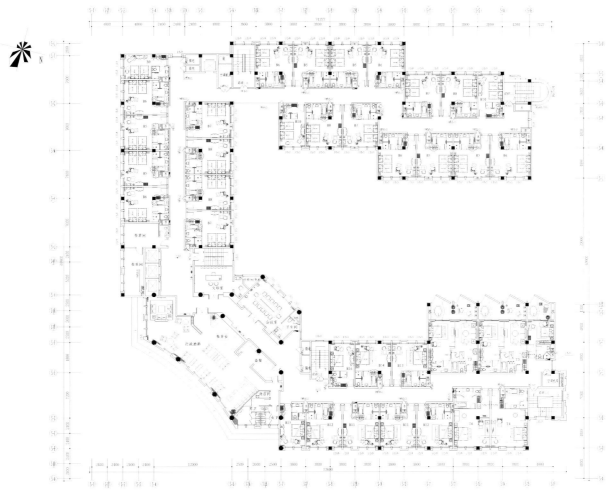

该案是按五星级标准投资兴建的"民国"建筑风格的温泉度假酒店，坐落于中国四大温泉之首的南京汤山。酒店分为南楼和北楼两栋建筑，南楼地上5层，局部6层，地下1层；北楼地上4层，地下2层。

酒店建筑的"民国"风格别树一帜，整体造型气势磅礴，彰显出酒店特有的尊贵和典雅。

步入大堂，三层高的中庭映入眼帘，米黄大理石、精美的铜花格柱子与内庭的景观相得益彰，令酒店更显洁净素雅与温馨，给人宾至如归的感觉。酒店拥有典雅舒适的各类客房164间/套，除了专设行政楼层与温泉套房外，3套不同装修风格的豪华别墅更是招待贵宾政要的首选。酒店中西餐厅、宴会包厢、特色餐厅、咖啡厅、高档酒吧等餐饮场地，为客人奉上各式珍馐美馔。酒店多功能厅和各类会议室，均配备先进的视听设备，能够满足不同规模、不同类型的会务接待需要。酒店温泉洗浴中心、健身房、网球场、迷你高尔夫、桌球室、KTV、棋牌室等都是放松身心、健康休闲的理想选择。散落在绿荫丛中、镶嵌在屋顶花园内的各式露天温泉汤池，更给人带来护肤与养生的诸多惊喜。

御至生活，尊享奢豪。酒店以客为尊、怡情山水的人文情怀将给海内外宾客带来清新的享受！

# Qingyuan Shampoola Resort Hotel

清远森波拉度假酒店

项目地点｜广东清远
项目面积｜总占地面积 666666 平方米
总建筑面积｜40000 平方米
设 计 公 司｜广州山晟旅游发展有限公司

Qingyuan Shampoola Resort Forest is composed of four parts, namely Shampoola Themed Park of Wonderful World, Shampoola Themed Hotel, Shampoola Volcanic Hot Spring, and Shampoola Glacier Water Canyon. The hotel is themed with ancient forests and Shampoola culture, with extensive use of ancient ship wood, volcanic stone, spinulose tree fern and other ancient elements and materials, and it is known as "five-star hotel emerge from jungle". The reception of the lobby is designed with ecological and plain ancient ship wood, coupled with antique stenciling rotating wooden doors and natural ebony root carvings. In functional design, the hotel supporting service area and the guestroom area are built independently to implement the "partition of the static and dynamic", fully meeting the functional requirements of leisurely vacation.

The Volcanic Hot Spring integrates the features of mountain hot spring and canyon hot spring, known as the pioneer of "the fifth-generation themed hot spring in China". It is divided into fragrance hot spring area, volcanic dynamic area and bamboo groves medicated bathing area, providing 78 originally ecological hot spring pools of different styles. The hundreds of meters long volcanic canyon lead guests enter into the ancient time space, where the 88-meter-high simulated volcanic re-produces the wonders of volcanic eruptions happened 80 million years ago, the scenery is so stirring!

The Glacier Water Canyon is created from the ruins of ancient glaciers in Fogang, and it is the first water park themed with glacier culture, featuring eight areas, they are glaciers tsunami surf area, moulin whirlpool drift area, glacier adventure rafting area, ice sky rainbow sliding area, Iceland Secrets Park of joy area, ice water stockade leisure area, glacial lake plaza culture area, indoor dressing and seating area.

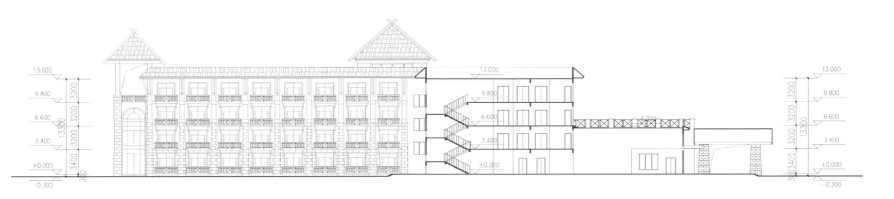

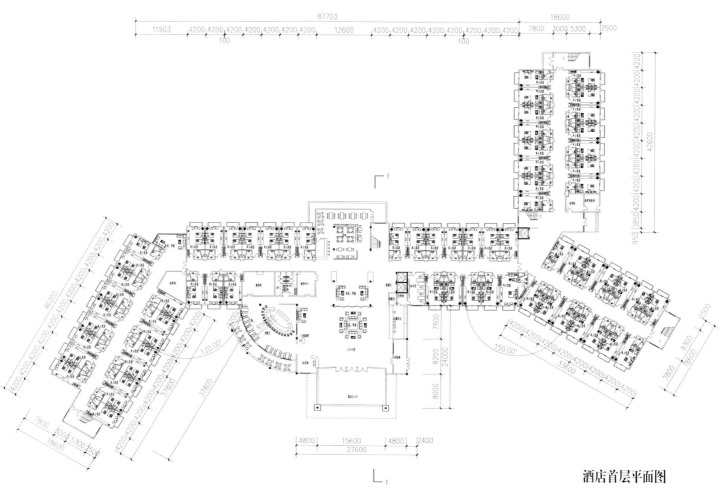

酒店首层平面图

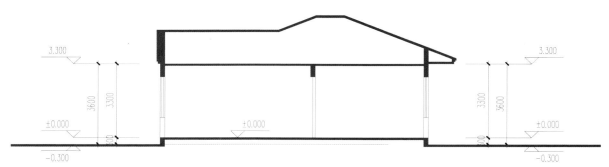

A型别墅1-1剖面图 1:100

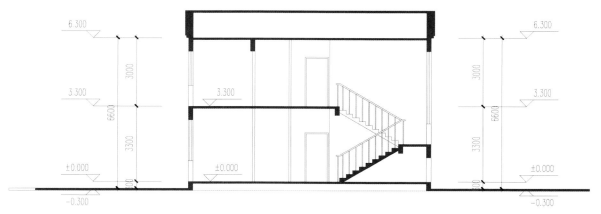

B型别墅2-2剖面图 1:100

清远森波拉度假森林集森波拉奇妙世界主题公园、森波拉主题酒店、森波拉火山温泉、森波拉冰川水谷四大板块于一体。酒店以远古森林和森波拉文化为主题，大量运用了古船木、火山石、桫椤树等远古元素和材料，被誉为"丛林里长出来的五星级酒店"。大堂接待前台采用生态古朴的古船木、古色古香的镂花旋转木门和纯天然的乌木根雕。在功能设置上，设计师将酒店配套服务区与客房区实行"动静分区"、独立建设，完全满足休闲度假的功能需求。

火山温泉集山地温泉和峡谷温泉特色于一体，被誉为"中国第五代主题温泉"的开创者。分为花香温泉区、火山动感区、竹林药浴区三大区域，拥有风格各异的78个原生态温泉浴池。百米长的火山峡谷引领人们进入远古时空，88米高的仿真火山复原了80万年前火山喷发的奇景，震撼人心！

冰川水谷建于佛冈古冰川遗址上，是广东首个以冰川文化为主题的亲水乐园，有冰川海啸冲浪区、冰臼漩涡漂移区、冰河历险漂流区、冰空彩虹滑道区、冰岛迷城欢乐区、冰花水寨休闲区、冰湖广场文化区、室内更衣休息区等八大区域。

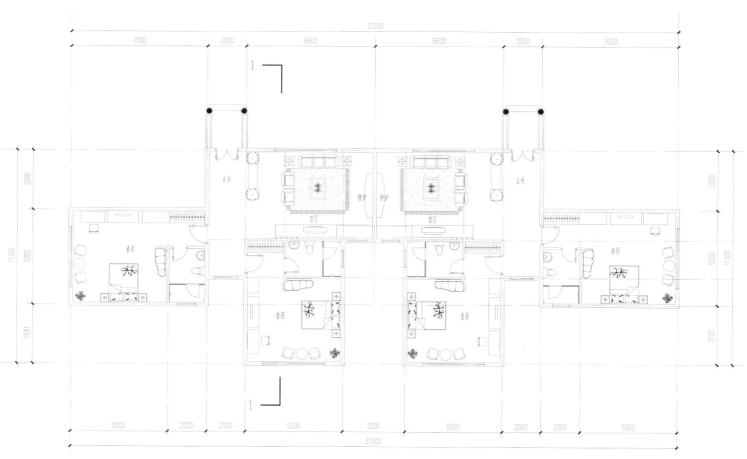

**A型别墅平面图** 1:100

**总: 240.6平方米**

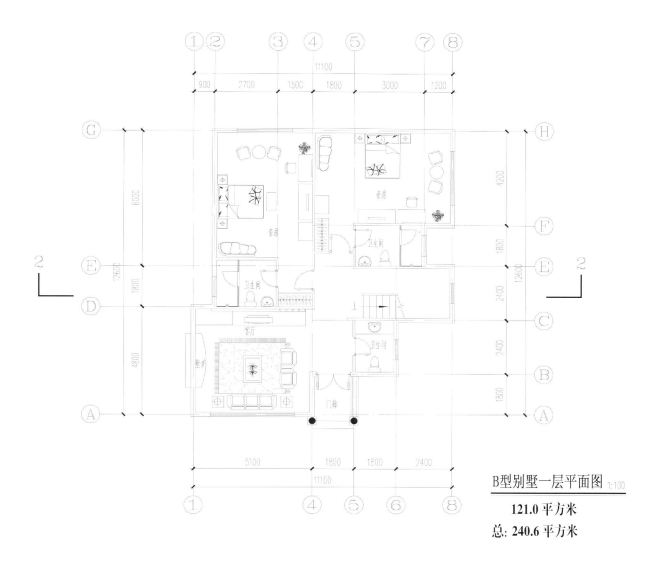

B型别墅一层平面图 1:100

**121.0 平方米**
**总: 240.6 平方米**

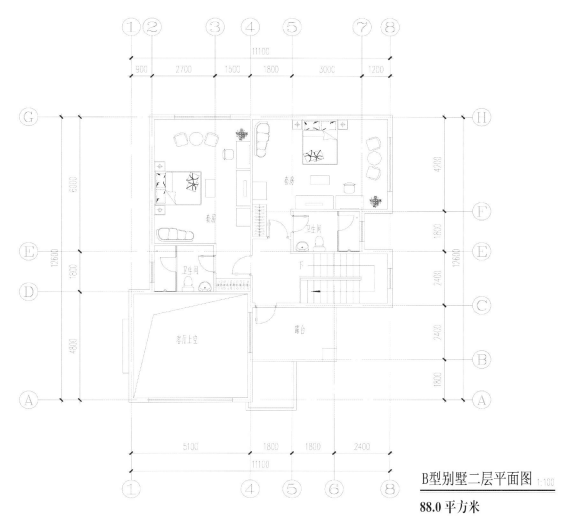

B型别墅二层平面图 1:100

**88.0 平方米**

267

# Nanchang Jinyan International Hot Spring Resort Island

南昌金燕国际温泉度假岛

**项 目 地 点** | 江西南昌
**项 目 面 积** | 总面积：1000000平方米；温泉池区面积：40000平方米
**温泉及景观设计** | 广州市山橡景观设计工程有限公司

The project is located in Shengmi Town, Xinjian County, Nanchang City, with only half an hour's drive from the urban center, it is the only large hot spring resort project in the circle of half an hour's drive in Nanchang City. The first phase of the project covers an area of about 100,000 square meters, including five-star hotel, apartments, hot spring area and forest bathing area. The second phase includes villas, townhouse, golf course and international school.

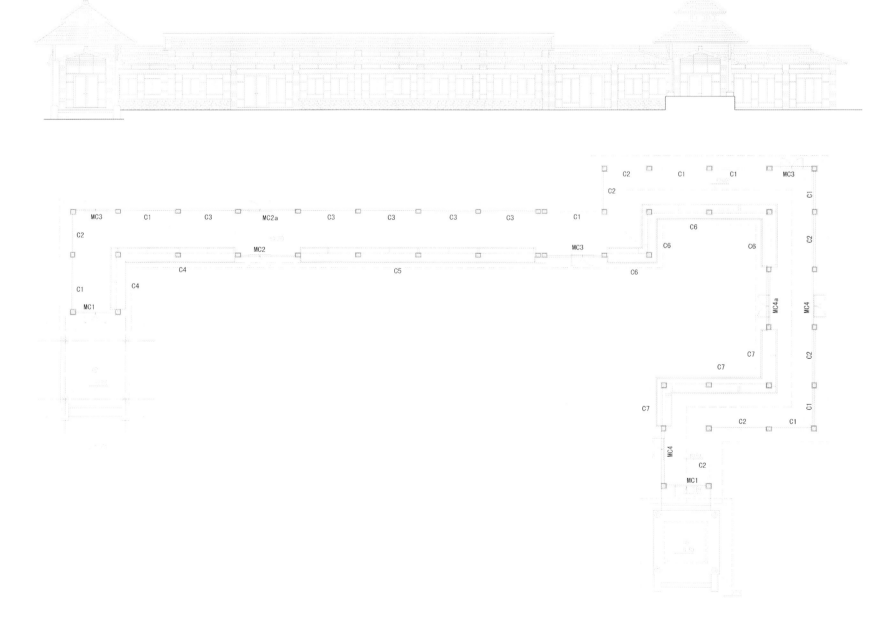

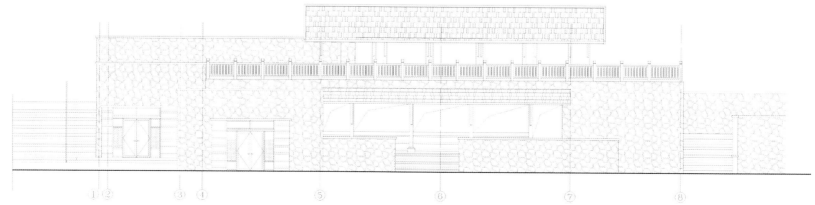

该案位于南昌市新建县生米镇,离中心市区只有半小时的车程,是南昌市半小时生活圈内唯一的大型温泉旅游度假项目。项目首期占地约100000平方米,包括五星级酒店、公寓、温泉区、森林汤屋区。二期规划包括别墅、洋房及高尔夫、国际学校等。

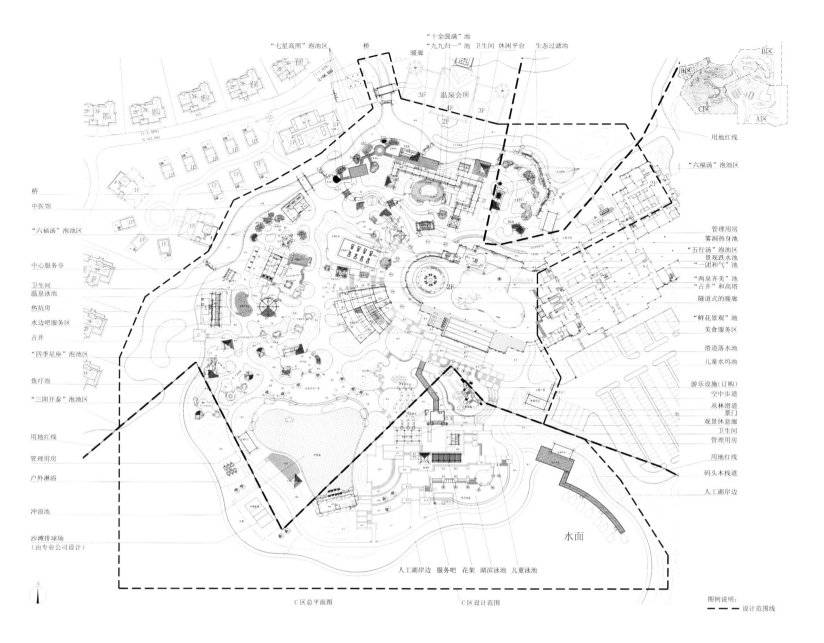

C区总平面图　　C区设计范围　　图例说明：
- - - 设计范围线

桥
中医馆
"六福汤"泡池区
中心服务亭
卫生间
温泉泳池
热炕房
水边吧服务区
占井
"四季星座"泡池区
鱼疗池
"三阳开泰"泡池区
用地红线
管理用房
户外淋浴
冲浪池
沙滩排球场
（由专业公司设计）

"七星高照"泡池区　桥　"十全圆满"池
暖廊　"九九归一"池　卫生间　休闲平台　生态过滤池
温泉会所

用地红线

"六福汤"泡池区
管理用房
雾洞热身池
"五行汤"泡池区
景观跌水池
"一团和气"池
"两泉齐美"池
"占井"和高塔
隧道式的暖廊
"鲜花景观"池
美食服务区
滑道落水池
儿童水坞池
游乐设施（订购）
空中步道
丛林滑道
景门
观景休息廊
卫生间
管理用房
用地红线
码头木栈道
人工湖岸边

水面

人工湖岸边　服务吧　花架　湖滨泳池　儿童泳池

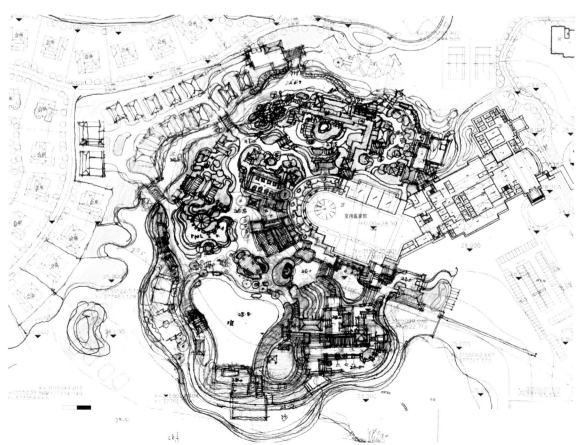

# Nanhu Travel Phoenix Lake International Hot Spring Resort

南湖国旅凤凰湖国际温泉度假村

| | |
|---|---|
| 项目地点 | 广东阳江 |
| 项目面积 | 4000000 平方米 |
| 设计公司 | 奥雅设计集团 |
| 客　　户 | 阳江市南湖国旅凤凰湖国际温泉度假村开发有限公司 |

### Design Vision

The designers take advantage of the local natural hot spring resources, committed to building a series of international five-star resort paradise with the theme of water, to drive the development of local tourism economy, and to show the natural ecology and local culture of Yangjiang.

### Design Goals

It is built to show the Asian Water Culture through water-related topics, and to build a focal point in Yangjiang, to show the local culture and boost the local tourism. In addition, it is a high-quality service center to entertain international guests, businessmen, executives and elites.

### Design Concept

With the theme of ecology and the designing language of Southeast Asian flavor, the project is developed to be a comprehensive resort. Phoenix Nirvana, ashes reborn, means that the development of the base is just like a phoenix transformed from pheasant. No matter the whole shape or function, it is a qualitative change. Its unique architectural form symbolizes the phoenix flying, responding to the resort theme.

### Design Strategy

The project features lush vegetation with landscape, and the most prominent point is the full use of the natural water resource to demonstrate the healthy and environmentally-friendly concept from Asian water culture, such as tea, organic food, spa and hot spring. The abundant water resources from the lake and small water bodies are used, with water as the dominant element of the overall plan, to form different water features, like lake, streams, pond, wetlands, and hot spring.

Creating different attractions in the same scenery is a distinctive selling point, reflecting the personality of the public space.

### Traffic Environmentally-friendly Concept

The project follows the principle of low-carbon transportation, introducing battery cars and bikes into the touring area, fully highlights the concept of environmental protection and low carbon.

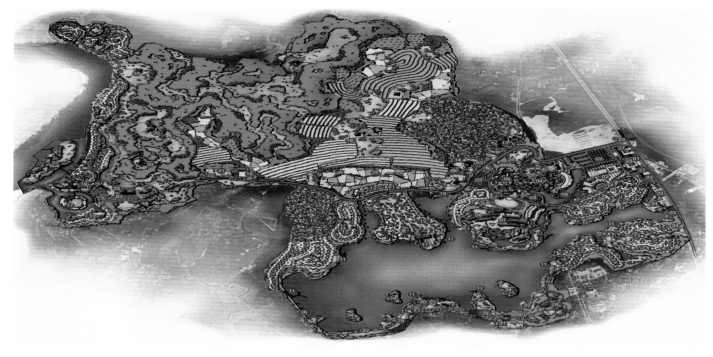

图例
01 主入口
02 温泉小镇
03 停车场
04 温泉休闲中心
05 园林度假酒店
06 儿童水上乐园
07 五星级酒店
08 国际会议中心
09 印象阳江
10 亲水湖堤
11 渡轮码头
12 低密度高尚住宅区
13 低密度高尚住宅区会所（红酒窖）
14 低密度高尚住宅区会所（网球场）
15 垃圾处理站
16 花卉植物观赏区
17 低密度高尚住宅区会所
18 开心农场会所咖啡厅
19 漠阳江景观游俱乐部会所酒店
20 养生馆（太极馆、瑜伽馆、氧气馆）
21 绿林养生体育公园
22 漠阳江生态岛
23 次入口（生态景观长廊）
24 游艇俱乐部
25 开心农场
26 亲水平台
27 入口广场
28 浪漫沙滩
29 植物观赏区
30 动物观赏区
31 污水处理站
32 绿色生态走廊
33 特色标志雕塑小品
34 生态小岛

**整体总平面图**
MASTER PLAN

景观湖　　浪漫沙滩　　游泳池　室外餐饮廊　副酒店会所　五层高的酒店主体

主题餐厅　　岗亭、服务站　室外休憩果餐区　岗亭、服务站　　一层高购物餐饮综合楼

公共温泉区　　公共温泉区　　独立温泉小屋　　休憩凉亭

场地剖面设计一
LONG SECTION Ⅰ

主酒店会所　贵宾车行道　生态种植区　酒店区内车行道　酒店入口岗亭　5米车行道　生态停车场　温泉休闲中心入口接待处

场地剖面设计二
LONG SECTION Ⅱ

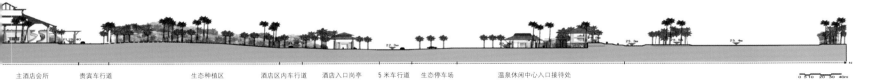

商业街内广场　　一层高购物餐饮综合楼　　商业街大广场　　温泉休闲中心入口接待处　　入口广场

场地剖面设计三
LONG SECTION Ⅲ

景观种植区　　　　　　　　　　　　温泉休闲中心入口接待处

### 设计愿景

设计师利用当地的天然温泉资源，致力于打造出一系列以水为主题的国际五星级度假休闲天堂，进而带动当地旅游经济的发展，展示阳江的自然生态与当地文化。

### 设计目标

通过与水相关的主题来体现亚洲水文化，将此地打造为阳江的一个亮点名片，展示阳江本土文化，带动当地旅游业发展。另外，也借此建设一个接待国际宾客、商人、行政人士等社会精英的高级优质服务中心。

### 设计理念

本案以生态为主题、以东南亚风情的设计语言，综合开发度假村。凤凰涅槃、浴火重生，暗喻基地的开发将如山鸡蜕变成凤凰一样，无论是整体的形态还是使用功能，都是一种质的改变。其独特的建筑形态，象征着凤凰飞舞，与度假村的主题风格相一致。

### 设计策略

本项目植被茂密，具有景观性，最突出的是充分利用天然的水资源来展现亚洲水文化中所蕴含的健康、环保理念，如饮茶、有机食品、水疗、温泉等。利用湖区和小水体所提供的丰富的水资源，把水作为总体规划的主导元素，形成湖、溪流、池塘、湿地、温泉等不同水景。

在同一景区中创造不同的景点作为各具特色的卖点，体现了公共空间中的个性。

### 交通环保理念

本案遵循低碳交通运作原则，在旅游区域内使用电瓶车、自行车，充分强调了环保低碳的理念。

# Riyuegǔ Hot Spring Country Club, Phase II

日月谷温泉乡村俱乐部二期

项目地点 | 福建厦门
项目面积 | 57000 平方米
设 计 师 | 赖连取、李德伟
景观设计 | HANCS Lazndscape Planning Japan Inc
发 展 商 | 坤城房地产开发有限公司
项目委托方 | 坤城房地产开发有限公司

Away from the busy city temporarily, saying goodbye to high-rising buildings, finding another happy world, you can relax mentally and physically in the green trees, gurgling spring, and enjoy the sun and moon – this is a journey feast of Southeast Asian style provided by Xiamen Riyuegǔ Hot Spring Country Club.

Riyuegǔ Hot Spring Villa features the advantages of leisure vacation and personalized homing, taking the functional requirements of the two different natures into account in the landscape, and integrating them into an organic whole, to fully show the elegant quality of the project. The club advocates the Lifestyle of Health and Sustainability, and the guestrooms introduce large area of log decorations, where the tables, chairs, beds and other supplies are made of natural rattan, straw, cotton, linen and other ecological materials, with the embellishment of vibrant plants, allowing guests to enjoy the happiness of deep contact and intimate connection with nature.

The facilities are exquisite and advanced without losing local flavor, and in the elegant and fresh environment, the looming background music like sounds of nature fiddles with people's thoughts. Here, you can swim in the blue swimming pool, drip with sweat in the sporty gym, and enjoy the professional services from the aromatherapist in the quiet and comfortable independent rooms, having a leisurely nap. Of course, you can also enjoy the unique features only in Riyuegǔ Hot Spring.

The designing style is mainly pastoral style. The elegant modern pastoral style pursues the sense of mind's natural belonging in designing, giving you a blowing of elegant fragrance. The open-style space structure, flowers and green plants everywhere, elegant fabric of various colors and patterns...all create a harmonious atmosphere from the whole, and form a new modern landscape.

总平面图

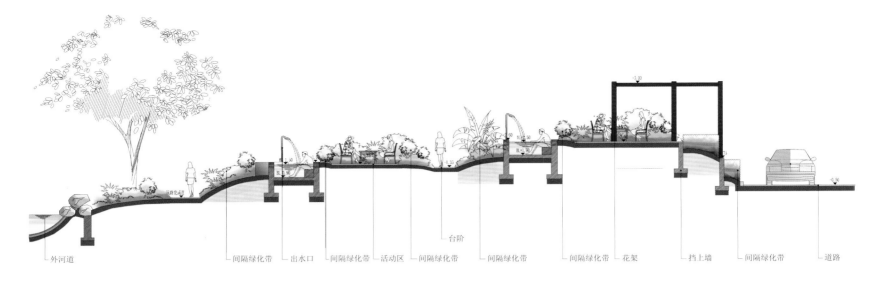

温泉区剖面详图一
SCALE 1:100

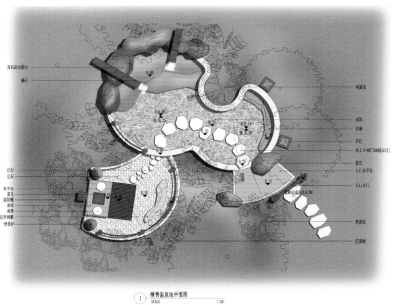

① 檀香温泉池平面图
SCALE 1:50

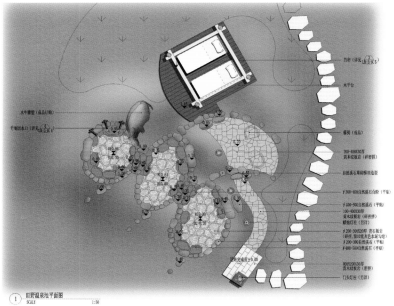

① 田野温泉池平面图
SCALE 1:50

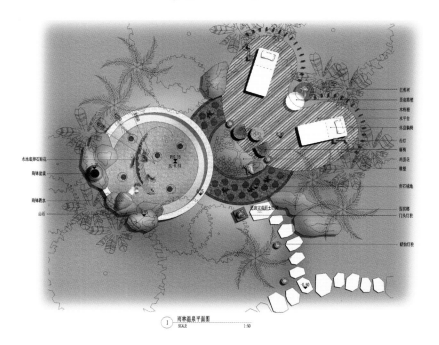

① 园林温泉平面图
SCALE 1:50

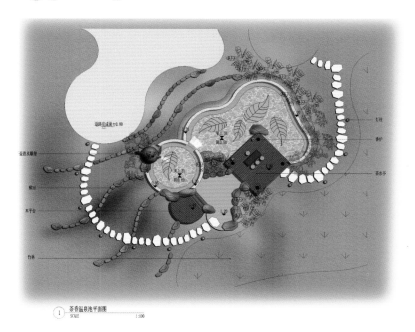

① 茶香温泉池平面图
SCALE 1:100

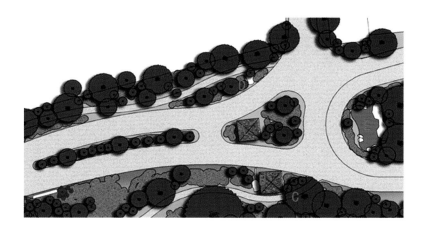
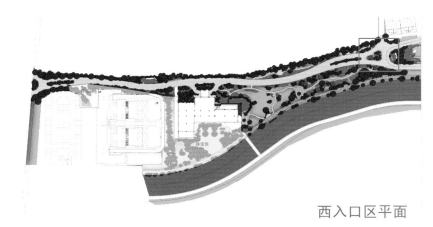

西入口区平面

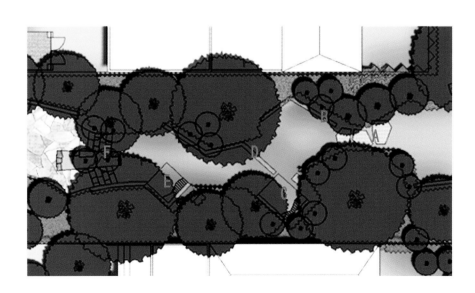
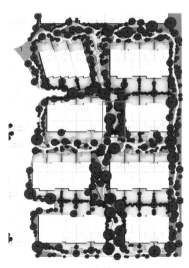

香草绿园平面

　　暂离繁忙都市，挥别高楼大厦，另辟乐活天地，让身心徜徉于幽幽翠绿、潺潺玉泉，涤荡于日月精华——这就是厦门日月谷温泉乡村俱乐部呈献给世人的东南亚风情休旅盛宴。

　　日月谷温泉别墅集休闲度假及个性居家住宅之所长，在景观上考虑了两者不同性质的功能需求，同时又将二者融合为一个有机整体，把项目的幽雅特质体现得淋漓尽致。倡导乐活概念的日月谷温泉度假村，客房内采用大面积的原木装饰，摆设的桌、椅、床饰等用品均由纯天然的藤、草绳、棉、麻等原生态材质制成，加上生机勃勃的植物点缀，让入住的宾客深切地体验到与自然亲密接触的喜悦。

　　设施既精致先进又不乏乡土气息，环境优雅清新、若隐若现的背景音乐天籁般拨弄人的思绪。在这里，既可畅游碧波泳池，也可以在动感十足的健身房挥汗如雨，更可以躺在安静舒适的独立芳疗室中，享受芳疗师提供的专业服务，悠然小憩。当然，你还可以享受只有在日月谷才能享受到的风格迥异的特色温泉。

　　在设计风格上，以田园式为主旨。雅趣的现代田园风格在设计上讲求心灵的自然归属感，给人一种扑面而来的淡雅芬芳。开放式的空间结构、随处可见的花卉绿植、各种花色的优雅布艺……一切都从整体上营造了一种和谐氛围、塑造出全新的现代景观。

# Yuanchang Yinhe Central Lake Hot Spring Resort

源昌银河湖心岛温泉度假村

项目地点｜福建泉州
项目面积｜30000 平方米
设 计 师｜赖连取、范晖
景观设计｜HANCS Landscape Planning Japan Inc
项目委托方｜南安源昌房地产开发有限公司

The project is located in Quanzhou. Quanzhou is also known as Licheng, Zaitun, Wenling, located in the southeast of Fujian province, facing Taiwan across the sea, and it is the staring point of the ancient "silk road on the sea". Quanzhou Port was known as "the largest oriental port" in Song and Yuan Dynasties, and Quanzhou is among the first 24 historical and cultural cities announced by the State Council, enjoying the reputation of "coastal land for cultural prosperity".

The marine style and personalized designing elements are the biggest selling points in the project. The style of each hot spring is different and unique. The design highlights the island style through the application and organization of various landscape elements, making the tropical island style, fresh building pattern and pavilions complement each other, to draw a natural picture integrating the tropical marine style, hot spring landscape and artificial ingenuity.

本项目位于泉州。泉州又称鲤城、刺桐城、温陵,地处福建东南部,与台湾地压隔海相望,是古代"海上丝绸之路"的起点。宋、元时期泉州港被誉为"东方第一大港",泉州也是国务院第一批公布的24个历史文化名城之一,素有"海滨邹鲁"的美誉。

# Huidong Baipenzhu Hot Spring Hotel

惠东白盆珠温泉酒店

**项目地点** | 广东惠州
**项目面积** | 总面积：174459 平方米
　　　　　　 净用地面积：105268 平方米
　　　　　　 总建筑面积：31000 平方米
**设 计 师** | 林学明、陈向京、曾芷君、齐胜利、徐婕媛、吴剑锋、刘鹰、谢云权、刘文静
**设计公司** | 广州集美组室内设计有限公司、北京东方华太建筑设计工程有限责任公司

The project is located on the south side of Pinggao highway, Baipenzhu Town, Huidong County, and the whole piece of land is basically mountainous, where the peak is on the southeast corner, extending to east, west and north. From north to south is 450 meters, and from east to west is 336 meters, with beautiful and clean surroundings, it is a perfect construction site for hot spring resort. The plan layout of the project is divided into public area and independent villas area, where in the public area, the hotel lobby is located on the top of the mountain with higher ground, enjoying a good view, and the guestroom buildings are scattered around the lobby on the slopes. The conference center, hot spring service center, commercial street and hot spring pool area are placed in gentle plain, close to the Pinggao highway, facilitating the flows of people and traffic. The independent villas area mainly includes landscape lounge, presidential villas, hot spring villas and detached villas, and the villas are interspersed and combined along the mountain. Since most of the engineering is mountain building, in the transport and vertical design, the designers take main consideration on the mountain to minimize earthworks. The roadway is composed mainly by 7m-wide main road and 4m-wide secondary road, and in vertical direction, each building monomer is designed on the slopes of mountain with staggered heights, without blocking the one in front, so they share the beautiful natural environment around.

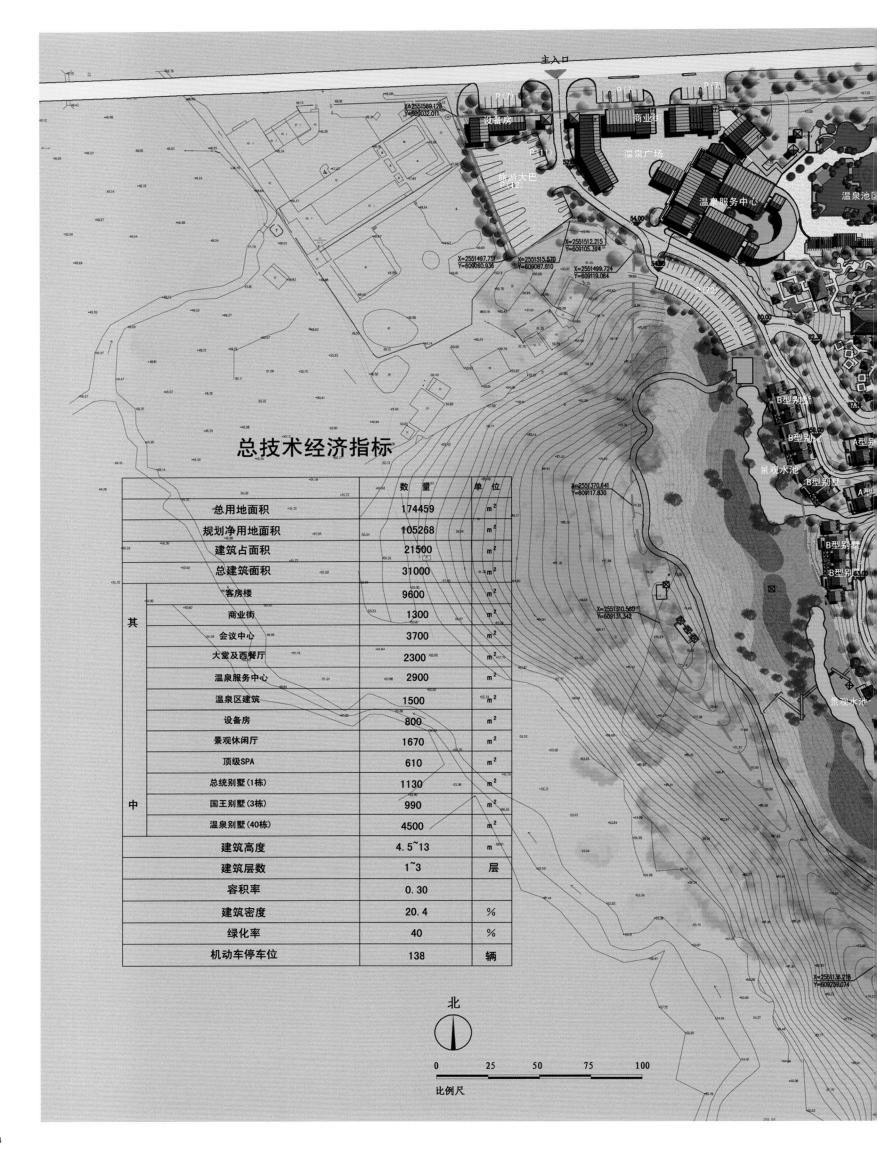

## 总技术经济指标

| | 项目 | 数量 | 单位 |
|---|---|---|---|
| | 总用地面积 | 174459 | m² |
| | 规划净用地面积 | 105268 | m² |
| | 建筑占地面积 | 21500 | m² |
| | 总建筑面积 | 31000 | m² |
| 其中 | 客房楼 | 9600 | m² |
| | 商业街 | 1300 | m² |
| | 会议中心 | 3700 | m² |
| | 大堂及西餐厅 | 2300 | m² |
| | 温泉服务中心 | 2900 | m² |
| | 温泉区建筑 | 1500 | m² |
| | 设备房 | 800 | m² |
| | 景观休闲厅 | 1670 | m² |
| | 顶级SPA | 610 | m² |
| | 总统别墅(1栋) | 1130 | m² |
| | 国王别墅(3栋) | 990 | m² |
| | 温泉别墅(40栋) | 4500 | m² |
| | 建筑高度 | 4.5~13 | m |
| | 建筑层数 | 1~3 | 层 |
| | 容积率 | 0.30 | |
| | 建筑密度 | 20.4 | % |
| | 绿化率 | 40 | % |
| | 机动车停车位 | 138 | 辆 |

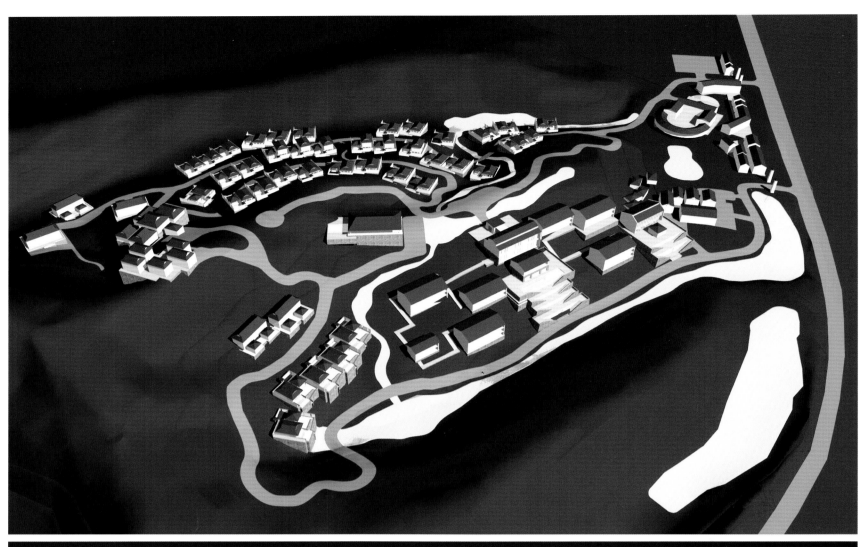
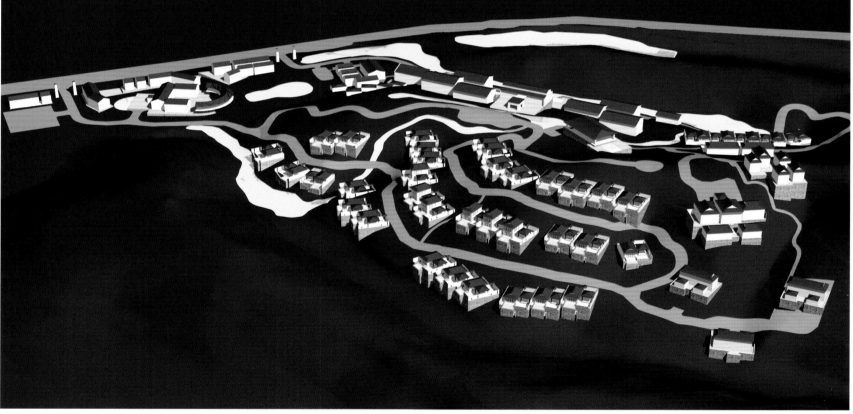

## 【商业及温泉服务中心首层平面图】

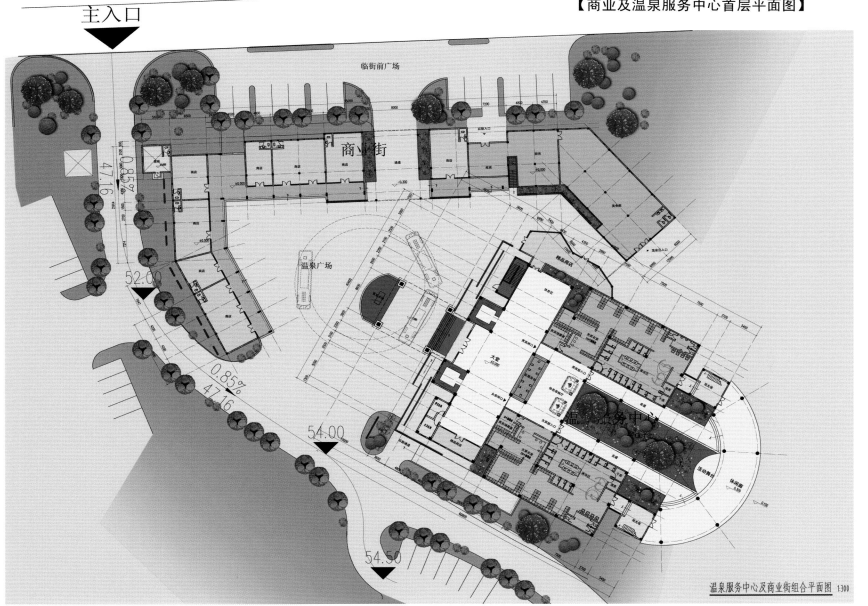

温泉服务中心及商业街组合平面图 1:300

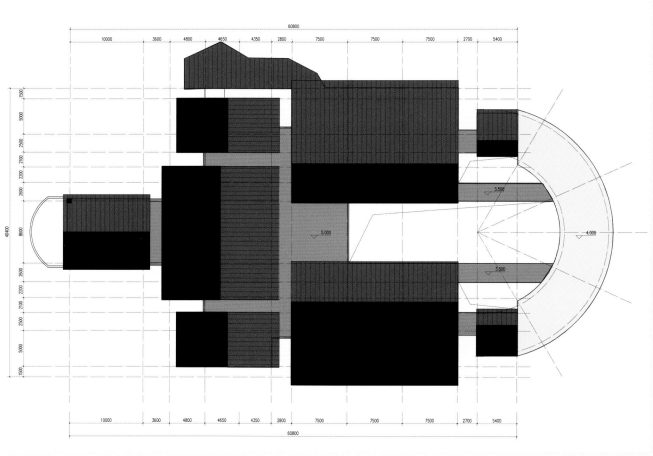

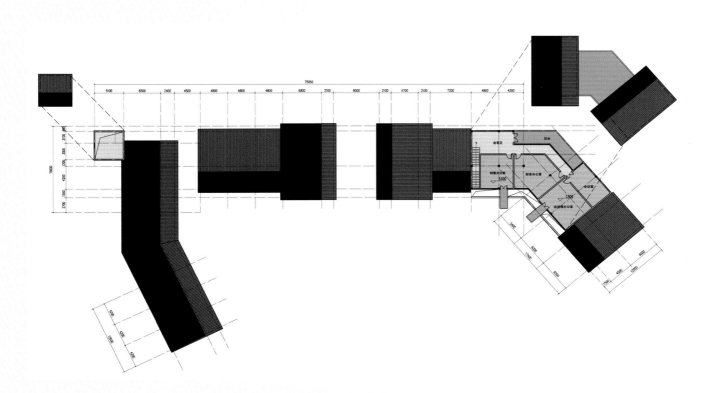

该案位于惠东县白盆珠镇平高公路南侧，整块用地基本为山地，东南角为山顶，向东、西、北几个方向延伸。其中南北方向长度为450米，东西方向为336米，四周环境优美整洁，是温泉度假酒店的良好建设用地。该项目的平面布局分为公共区域和独立别墅区，在公共区域中酒店大堂位于地势较高的山顶上，可享受良好的视野，客房楼顺应山势散落在大堂周边的坡地上。会议中心、温泉服务中心、商业街及温泉池区则布置在地势较缓的平地上，靠近平高公路，方便人流及车流的畅通。独立别墅区主要包括：景观休闲厅、总统别墅、温泉别墅及独栋别墅，各栋建筑依据山势穿插组合。由于本工程大部分为山地建筑，在交通及竖向设计上，主要考虑依据山势，尽量减少土方工程。车行道主要由7m宽的主干道及4m宽的次干道组成，竖向方向各栋建筑单体依据山势、层层错开高差，使每栋建筑单体都不会受到前一栋单体的阻挡，可以共享周边优美的自然环境。

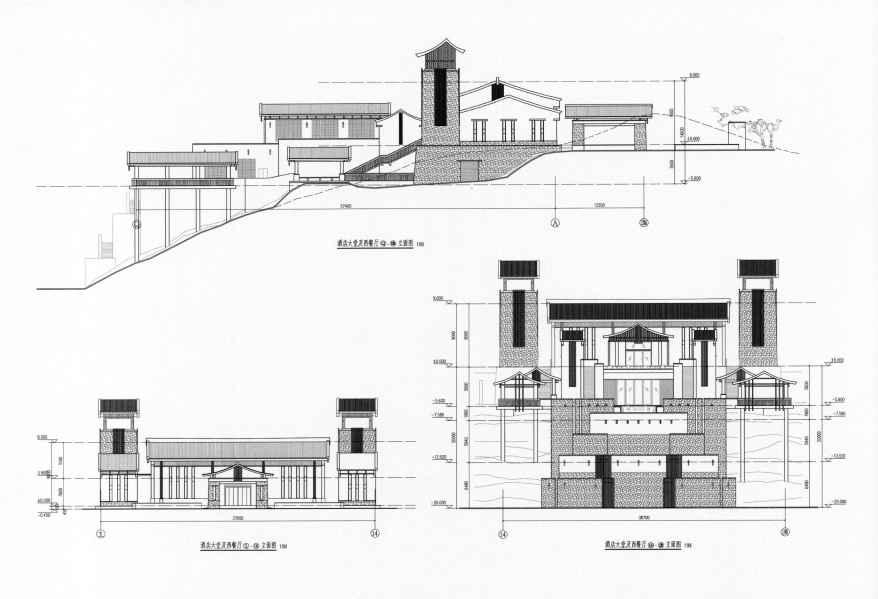
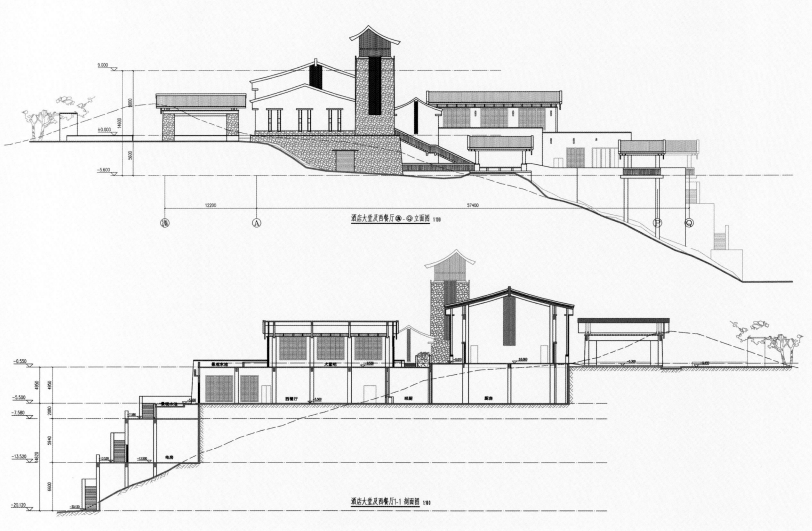

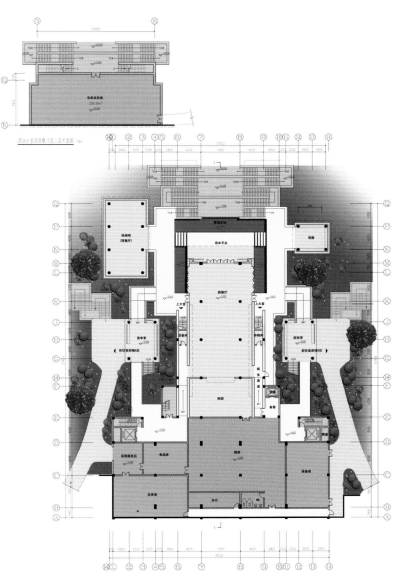

**本层建筑面积：1294.56 平方米**

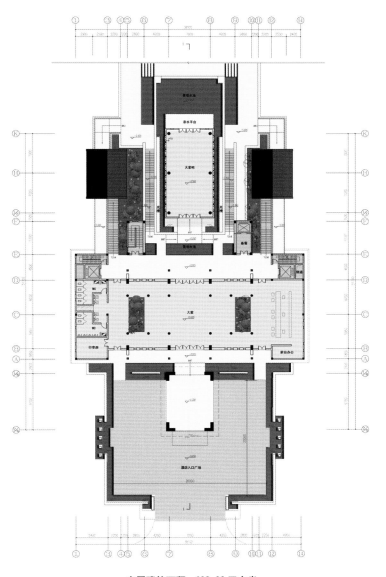

**本层建筑面积：628.22 平方米**
**总建筑面积：1922.78 平方米**

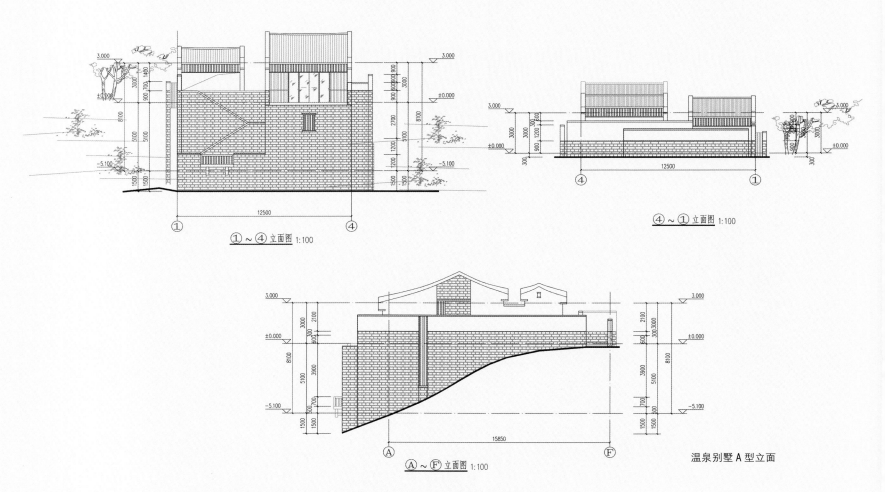

①~④ 立面图 1:100

④~① 立面图 1:100

Ⓐ~Ⓕ 立面图 1:100

温泉别墅 A 型立面

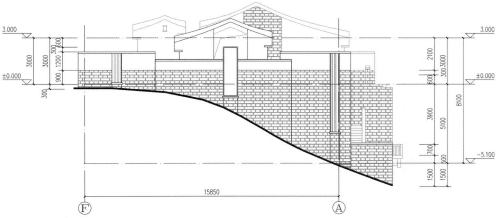

F～A 立面图 1:100

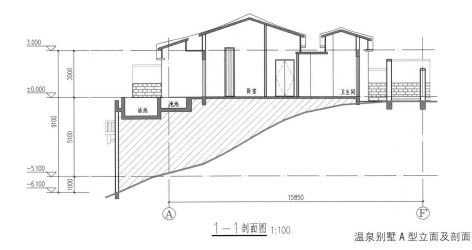

1—1 剖面图 1:100

温泉别墅 A 型立面及剖面

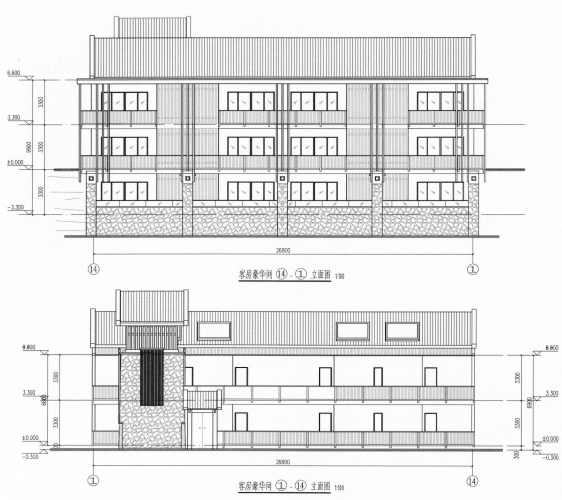

客房豪华间 ⑭ - ① 立面图 1:100

客房豪华间 ① - ⑭ 立面图 1:100

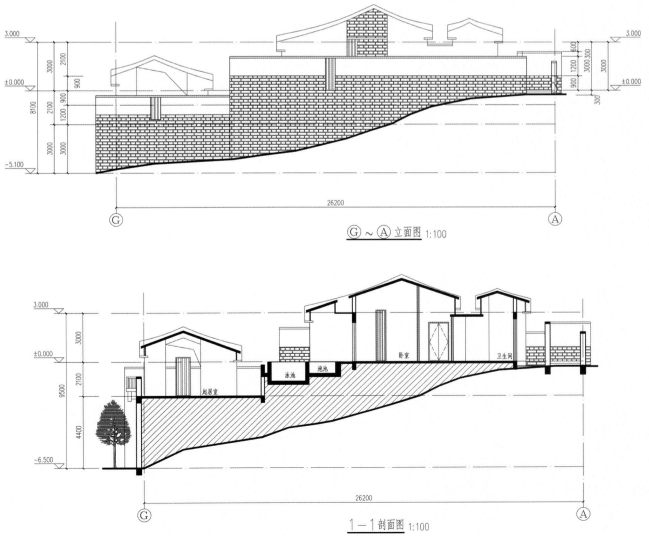

G～A 立面图 1:100

1—1 剖面图 1:100

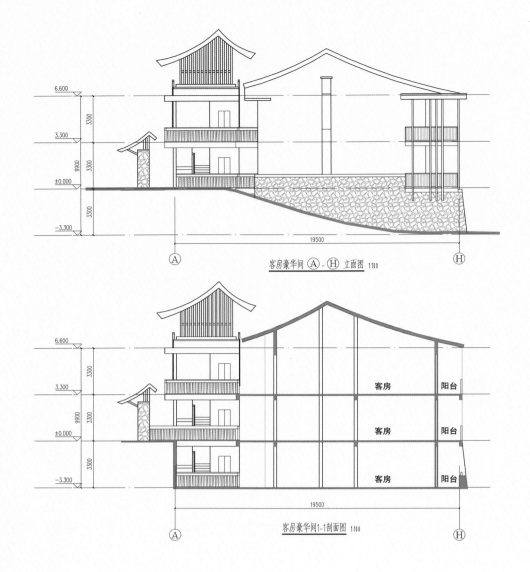

客房豪华间 Ⓐ-Ⓗ 立面图 1:100

客房豪华间 1-1 剖面图 1:100

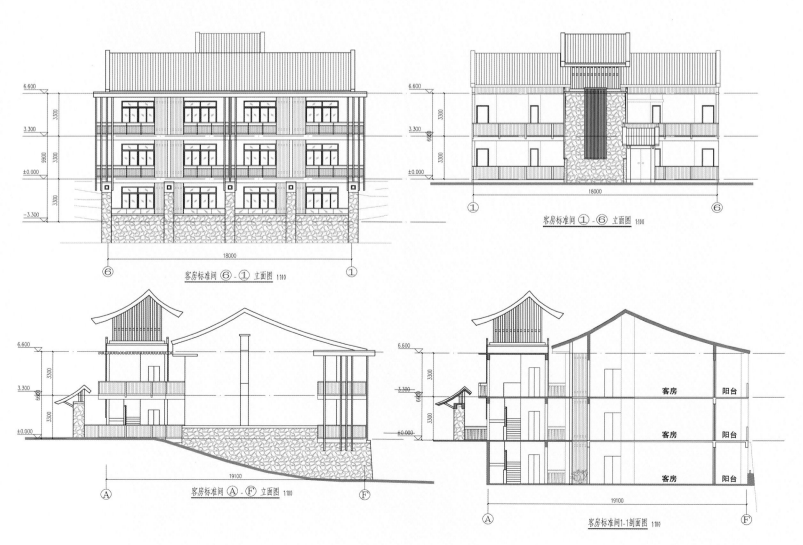

客房标准间 ⑥-① 立面图 1:100

客房标准间 ①-⑥ 立面图 1:100

客房标准间 Ⓐ-Ⓕ 立面图 1:100

客房标准间 1-1 剖面图 1:100

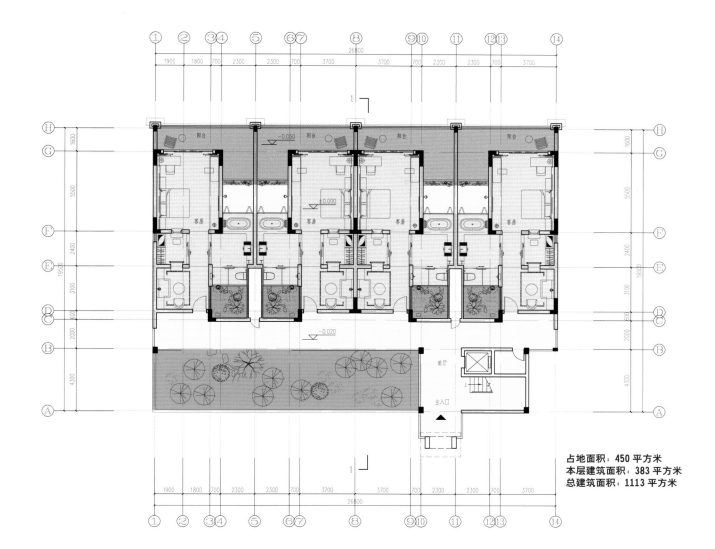

占地面积：450 平方米
本层建筑面积：383 平方米
总建筑面积：1113 平方米

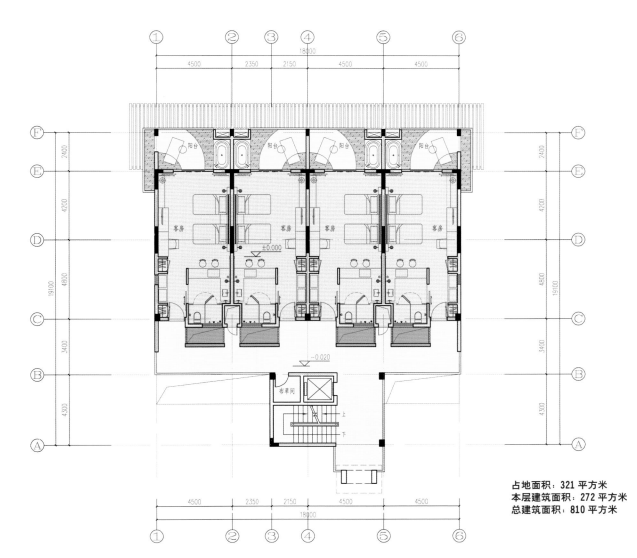

占地面积：321 平方米
本层建筑面积：272 平方米
总建筑面积：810 平方米

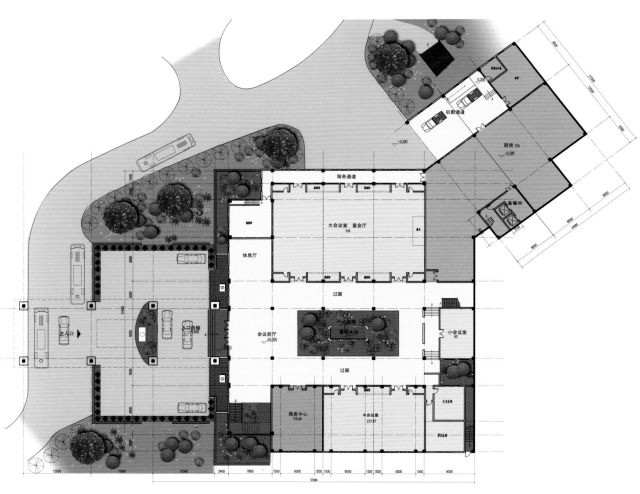

**本层建筑面积：2271 平方米**
**总建筑面积：3704 平方米**

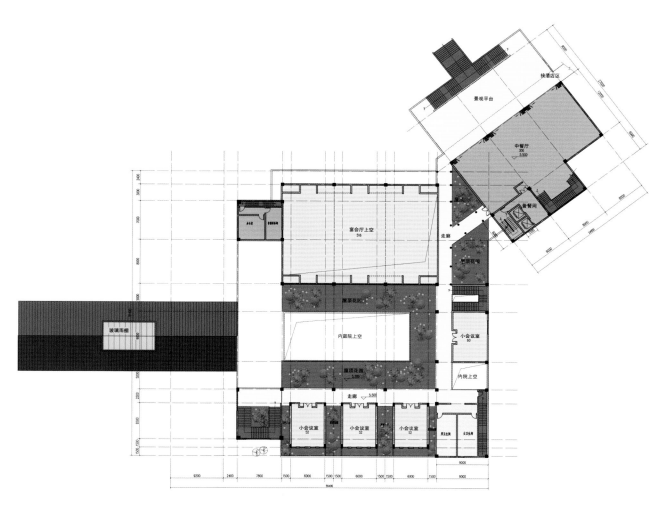

**本层建筑面积：963 平方米**

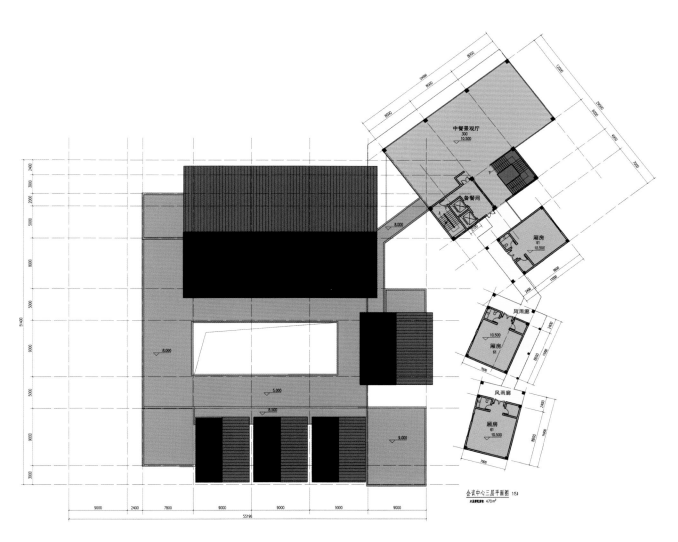

**本层建筑面积：470 平方米**

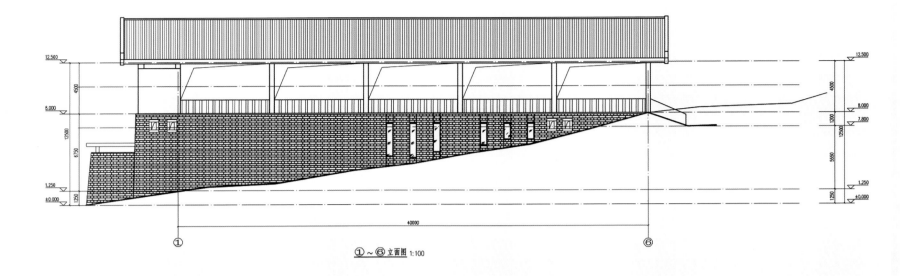

①~⑥ 立面图 1:100

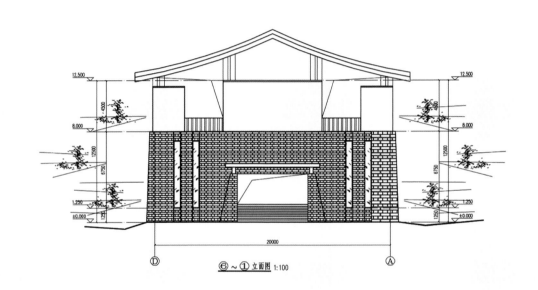

⑥~① 立面图 1:100

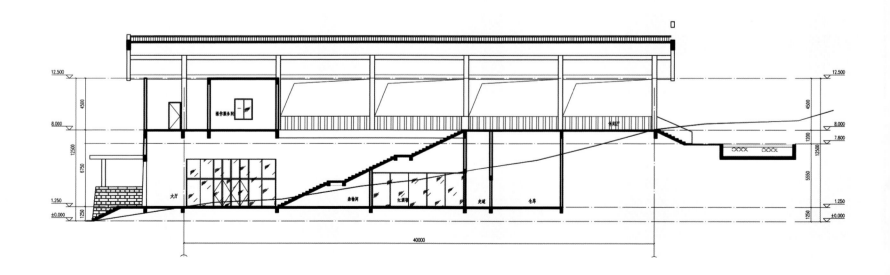

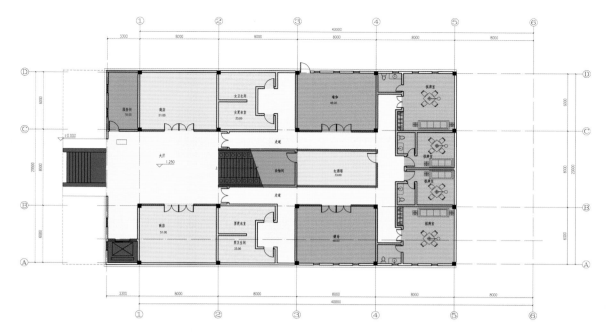

**本层建筑面积：730 平方米**
**总建筑面积：1600 平方米**

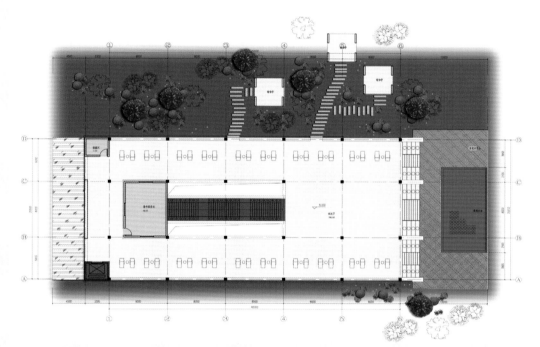

**本层建筑面积：870 平方米**

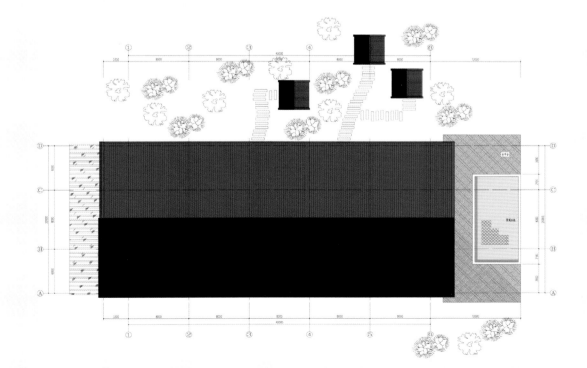

# INDEX 索引

**Alberto Apostoli**

Alberto Apostoli 出生于意大利、维罗那，毕业于威尼斯大学。尤其擅长设计住宅、商业建筑、酒店与渡假村、Spa 及保健中心、零售店及商铺、展览设计及办公空间（陈列室）以及产品的设计。公司实体包括 25 位专业资深的建筑师、MEP 工程师、室内以及工业设计师等。

**广州山晟旅游发展有限公司**

公司秉承"山外有山、不断超越"的文化理念，发挥旅游研究、规划设计与实操落地相结合的优势，提供旅游项目策划、旅游总体规划、修建性详细规划、旅游景观设计、创意建筑设计等方面的旅游项目整体服务。

公司网址：www.gzssly.com

**孙彦清**

金螳螂建筑装饰股份有限公司，金螳螂第十八设计院（副院长），中国建筑学会室内设计分会员。2007~2008 年度十大设计师。主创空间：酒店、会所；主要获奖：2007 年金羊奖（南京世茂滨江会所），2008 年金羊奖（南京金鼎湾私家会所），2011 年中国装饰奖铜奖酒店类，被评为国优工程（南京御豪汤山温泉度假酒店）。

洲联集团凭借欧洲顶级高科技生态节能技术、中国甲级设计资质、丰富的大型工程施工图经验以及优秀的研究实力和创意表现，率先在国内提出"5+1"一站式全程服务模式。"5+1"整合城市规划、建筑设计、景观设计、室内设计、平面设计五大类专业技术，外加市场研究、产品策划增值及工程设计总承包服务，为城市开发和地产行业提供全产业链的技术与顾问服务。

公司网址：www.www5a.com

**澳大利亚 SDG 设计集团**

设计产品的社会性和技术性成功是 SDG 设计团队一直追求的目标。大众的认可是设计产品得以成功之本；而建筑只有融合了各种先进技术才具有时代性，才能更好地服务于人类、服务于社会。社会性和技术性完美统一永远是 SDG 创造力的源泉和前进的动力。

公司网址：www.sdg-cn.com

**郭文河**

广州文和铧美园林集团总设计师、中国高级环境艺术师、中国建设部环艺会理事、《时代楼盘》等多家媒体及协会特聘顾问；曾获"中国人居最佳景观设计方案金奖""青岛市优秀建设工程设计创作大奖"等；代表作：北京保利垄上别墅、珠海方圆明月山溪、大连中信海港城等。

**薛峰**

深圳市寅界建筑室内设计有限公司，毕业于中央工艺美术学院，中国设计业"十大杰出青年"，中国高级室内建筑师，中国建筑装饰协会专家委员会专家。

**上海胜异设计顾问有限公司**

公司由姚胜虎总监初创于 1999 年，2011 年正式在上海成立胜异设计顾问有限公司。倡导策划先行，以"回归本质，崇尚人文与自然密切相融"为设计理念，结合科学的艺术表现手法，诠释多元化的生活方式，并坚持不懈地为客户提供最具价值和生命力的原创设计臻品。

**城市设计联盟**

公司运作至今，在实务实践的过程里，一直秉持着本身对设计的热诚及理念。设计师期待自己在设计上不只是思考如何找出有趣且前卫的概念来发展，还要关注建筑与周遭环境的关系及在都市发展上所能扮演的角色。思考建筑如何回应环境在地景、区域、城市甚至全球地域范围内带来的影响和冲击，及如何藉由设计引发对未来人类生活可能关心的议题。

**ACLA Limited**

ACLA（傲林国际）是一家享誉国际的设计咨询公司，拥有超过 30 年的历史，主要从事景观设计、总体规划、城市设计、旅游规划、建筑设计等多项领域的设计咨询工作。公司设址于香港、北京、上海、重庆、越南河内及胡志明市。在亚洲和中东地区承担了许多具有标志性的重点项目，为城市环境改善和景观发展做出了突出贡献，并屡获国际奖项及殊荣。

公司网址：www.acla.com.hk

**北京维拓时代建筑设计有限公司**

公司始建于1979年，具有建筑甲级及轻纺行业甲级设计资质，兼有工程咨询甲级及工程总承包资质，并于2001年通过ISO9000质量体系认证。秉承"换位、沟通、创新、求精"的设计方针，不断实现产品的设计创新与价值共赢。遵循"追求一流设计、确保质量第一"的质量方针，精心完成每一个作品。
公司网址：www.vtjz.com

**龚小刚**

毕业于天津工艺美术学院，现任北京龚氏建筑设计有限公司董事长。1980年开始从事家居设计工作，多次获奖；1993年创建个人设计工作室；2006年陕西西安临潼华清爱琴海国际温泉酒店室内设计；2008年北京首地·大峡谷商场室内设计；2009年航天部北京永丰基地五院、九院室内设计；2011年内蒙古呼和浩特甲35号五星会所建筑和室内设计；2012年北京天雅木樨园女装大厦室内设计。

**集美组**

成立于1994年，由加拿大SHERMAN DESIGN AND CONSTRUCTION与广州集美组设计公司合资建立，拥有有海外教育背景的专业人士组成的强大的设计策划及工程管理团队。经过近二十年的发展，已经成为在中国室内设计工程领域内享有盛誉的机构，为业主提供广泛的室内设计咨询以及工程服务。
公司网址：www.newsdays.com.cn

**厦门艺道景观规划设计有限公司**

公司主要从事主题旅游度假区温泉景观规划设计、公园景观规划、城市设计规划、居住区、城市广场、商业区域、市政园林等一系列的户外项目建设。作品既体现了西方的设计理念，又融合了东方的审美情趣，在为业主服务之外，更注重人与自然景观的营造，为新一轮的城市建设发挥自身的能力。
公司网址：www.alandplan.com

**Fabian Architects**

自1974年于开普敦成立以来，公司无论在概念设计、设计开发、产品文档、项目管理还是其他建筑环境相关的领域，都获得了颇丰的专业知识和丰富的实践经验。公司已为来自开普敦、约翰尼斯堡、伦敦等地的个人或机构房产投资公司和开发公司设计了众多项目。设计和产品文档部门提供高质的服务，该部门拥有15位建筑师、5位室内设计师、15位资深制图员以及由一批高效的执政及文书人员组成的技术娴熟的CAD部门。任何一个项目的设计师都会与客户协商合作，充分发挥团队能力。他们的热情、信誉、专业使公司在开普敦的建筑设计领域中发挥着积极且重要的作用。

**Mario Botta**

Mario Botta 于1943年出生于Mendrisio，Ticino。在Lugano经过一段时期的学徒之后，他进入米兰艺术学院学习，且同时在威尼斯建筑学会进修。在Carlo Scarpa 和 Giuseppe Mazzariol 的指导下，于1969年获得了他的职业称号。他的职业生涯于1970年始于Lugano，在他Ticino的独立住所建成以后，他开始探索所有的建筑类型：学校、银行、政府机构、图书馆、博物馆和宗教建筑。他将所有精力都用于建筑领域，是1996年Mendrisio建筑学会的发起者，2011至2013年仍然管理、执教于Mendrisio建筑学会。他的作品在许多展览中获奖且屡被展示。

**瀚世景观规划设计有限公司**

HANCS LANDSCAPE PLANNING CO.,LTD 成立于1973年，是一所大型跨国设计公司，在中国、中国台北、马来西亚、法国、日本北海道等都设有分支机构，拥有一支国际化、综合化的协作团队。HANCS设计团队重视每一个设计细节，秉承将人本、工艺、自然、文化、尺度相结合的理念，同时高度关注客户需求，把为客户创造最大价值作为第一目标，三十多年来致力于景观设计的创新与探索，设计作品遍布世界各地。

**广州市山橡景观设计工程有限公司**

广州市山橡景观设计工程有限公司致力于旅游区规划设计、温泉度假酒店园林设计、城市景观设计及居住区园林设计。山橡景观主创成员具有服务合生创展、香江集团、美林基业集团、侨鑫集团、珠江地产、正域集团的经历，设计作品在中国获得各类地产商业奖项。凭借在地产景观中积累的丰富经验，山橡公司进入了温泉酒店领域，相继打造了"南昆山锦绣香江温泉城""中山温泉宾馆""南昌金燕温泉度假岛""花都美林湖温泉酒店"等多个高端的温泉酒店项目，获得业界的一致赞誉。
公司网址：www.shanxiangjg.com

**奥雅设计集团介绍**

奥雅设计集团于1999年由李宝章先生在香港创立，前身为奥雅园境师事务所。目前在深圳、上海、北京、西安、青岛共有五个公司，拥有来自不同文化和学术背景的各类设计人员近四百人，是国内规模最大的、综合实力最强的景观规划设计公司之一。

奥雅集团致力于为中国城市化的发展提供从用地分析、经济策划、城市规划、建筑设计、景观设计和生态技术咨询等服务和全程化、一体化和专业化的解决方案，以创造具有地域特色的、人性化和充满活力的城市和城市空间。

**广州市品祺装饰设计工程有限公司**

2007年第四届国际室内设计大奖分别获得会所工程设计金奖、商业工程设计银奖、餐馆/酒吧工程设计银奖、酒店工程设计铜奖，2008年中国最强内设计调查（TOP100）当选企业，2008年金羊奖"中国十大商业空间设计师"，2008年第七届国际室内设计双年展优秀奖，2009年金羊奖"中国百杰室内设计师"，2009年金堂奖中国饭店业设计，2010年金堂奖优秀酒店空间设计，2010年金堂奖十佳娱乐空间设计，2010年金意陶杯广大省十大新锐作品三等奖，2010年广州国际园林景观与美好人居博览会优秀作品奖。